A History of Aeronautics

E. Charles Vivian

Contents

FOREWORD 7
PART I. THE EVOLUTION OF THE AEROPLANE 9
   I. THE PERIOD OF LEGEND 9
   II. EARLY EXPERIMENTS 16
   III. SIR GEORGE CAYLEY--THOMAS WALKER 34
   IV. THE MIDDLE NINETEENTH CENTURY 42
   V. WENHAM, LE BRIS, AND SOME OTHERS 52
   VI. THE AGE OF THE GIANTS 59
   VII. LILIENTHAL AND PILCHER 67
   VIII. AMERICAN GLIDING EXPERIMENTS 75
   IX. NOT PROVEN 84
   X. SAMUEL PIERPOINT LANGLEY 91
   XI. THE WRIGHT BROTHERS 99
   XII. THE FIRST YEARS OF CONQUEST 119
   XIII. FIRST FLIERS IN ENGLAND 127
   XIV. RHEIMS, AND AFTER 134
   XV. THE CHANNEL CROSSING 141
   XVI. LONDON TO MANCHESTER 145
   XVII. A SUMMARY, TO 1911 148
   XVIII. A SUMMARY, TO 1914 156
   XIX. THE WAR PERIOD--I 164
   XX. THE WAR PERIOD--II 172
   XXI. RECONSTRUCTION 176
   XXII. 1919-20 179
PART II. 1903-1920: PROGRESS IN DESIGN By Lieut.-Col. W. Lockwood Marsh 184
   I. THE BEGINNINGS 184
   II. MULTIPLICITY OF IDEAS 192
   III. PROGRESS ON STANDARDISED LINES 197
   IV. THE WAR PERIOD 203
PART III. AEROSTATICS 209
   I. BEGINNINGS 209
   II. THE FIRST DIRIGIBLES 218
   III. SANTOS-DUMONT 225
   IV. THE MILITARY DIRIGIBLE 229
   V. BRITISH AIRSHIP DESIGN 236
   VI. THE AIRSHIP COMMERCIALLY 244
   VII. KITE BALLOONS 247
PART IV. ENGINE DEVELOPMENT 250
   I. THE VERTICAL TYPE 250
   II. THE VEE TYPE 262
   III. THE RADIAL TYPE 270
   IV. THE ROTARY TYPE 276
   V. THE HORIZONTALLY-OPPOSED ENGINE 282
   VI. THE TWO-STROKE CYCLE ENGINE 286
   VII. ENGINES OF THE WAR PERIOD 293

APPENDIX A GENERAL MENSIER'S REPORT ON THE TRIALS OF CLEMENT ADER'S AVION. 300
APPENDIX B 307
APPENDIX C 321

A HISTORY OF AERONAUTICS

BY
E. Charles Vivian

FOREWORD

Although successful heavier-than-air flight is less than two decades old, and successful dirigible propulsion antedates it by a very short period, the mass of experiment and accomplishment renders any one-volume history of the subject a matter of selection. In addition to the restrictions imposed by space limits, the material for compilation is fragmentary, and, in many cases, scattered through periodical and other publications. Hitherto, there has been no attempt at furnishing a detailed account of how the aeroplane and the dirigible of to-day came to being, but each author who has treated the subject has devoted his attention to some special phase or section. The principal exception to this rule--Hildebrandt--wrote in 1906, and a good many of his statements are inaccurate, especially with regard to heavier-than-air experiment.

Such statements as are made in this work are, where possible, given with acknowledgment to the authorities on which they rest. Further acknowledgment is due to Lieut.-Col. Lockwood Marsh, not only for the section on aeroplane development which he has contributed to the work, but also for his kindly assistance and advice in connection with the section on aerostation. The author's thanks are also due to the Royal Aeronautical Society for free access to its valuable library of aeronautical literature, and to Mr A. Vincent Clarke for permission to make use of his notes on the development of the aero engine.

In this work is no claim to originality--it has been a matter mainly of compilation, and some stories, notably those of the Wright Brothers and of Santos Dumont, are better told in the words of the men themselves than any third party could tell them. The author claims, however, that this is the first attempt at recording the facts of development and stating, as fully as is possible in the compass of a single volume, how flight and aerostation have evolved. The time for a critical history of the subject is not yet.

In the matter of illustrations, it has been found very difficult to secure suitable material. Even the official series of photographs of aeroplanes in the war period is curiously incomplete' and the methods of censorship during that period prevented any complete series being privately collected. Omissions in this respect will probably be remedied in future editions of the work, as fresh material is constantly being located.

E.C.V. October, 1920.

PART I. THE EVOLUTION OF THE AEROPLANE
I. THE PERIOD OF LEGEND

The blending of fact and fancy which men call legend reached its fullest and richest expression in the golden age of Greece, and thus it is to Greek mythology that one must turn for the best form of any legend which foreshadows history. Yet the prevalence of legends regarding flight, existing in the records of practically every race, shows that this form of transit was a dream of many peoples--man always wanted to fly, and imagined means of flight.

In this age of steel, a very great part of the inventive genius of man has gone into devices intended to facilitate transport, both of men and goods, and the growth of civilisation is in reality the facilitation of transit, improvement of the means of communication. He was a genius who first hoisted a sail on a boat and saved the labour of rowing; equally, he who first harnessed ox or dog or horse to a wheeled vehicle was a genius--and these looked up, as men have looked up from the earliest days of all, seeing that the birds had solved the problem of transit far more completely than themselves. So it must have appeared, and there is no age in history in which some dreamers have not dreamed of the conquest of the air; if the caveman had left records, these would without doubt have showed that he, too, dreamed this dream. His main aim, probably, was self-preservation; when the dinosaur looked round the corner, the prehistoric bird got out of the way in his usual manner, and prehistoric man, such of him as succeeded in getting out of the way after his fashion--naturally envied the bird, and concluded that as lord of creation in a doubtful sort of way he ought to have equal facilities. He may have tried, like Simon the Magician, and other early experimenters, to improvise those facilities; assuming that he did, there is the groundwork of much of the older legend with regard to men who flew, since, when history began, legends would be fashioned out of attempts and even the desire to fly, these being compounded of some small ingredient of truth and much exaggeration and addition.

In a study of the first beginnings of the art, it is worth while to mention even the earliest of the legends and traditions, for they show the trend of men's minds and the constancy of this dream that has become reality in the twentieth century. In one of the oldest records of the world, the Indian classic Mahabarata, it is stated that 'Krishna's enemies sought the aid of the demons, who built an aerial chariot with sides of iron and clad with wings. The chariot was driven through the sky till it stood over Dwarakha, where Krishna's followers dwelt, and from there it hurled down upon the city missiles that destroyed everything on which they fell.' Here is pure fable, not legend, but still a curious forecast of twentieth century bombs from a rigid dirigible. It is to be noted in this case, as in many, that the power to fly was an attribute of evil, not of good--it was the demons who built the chariot, even as at Friedrichshavn. Mediaeval legend in nearly every case, attributes flight to the aid of evil powers, and incites well-disposed people to stick to the solid earth--though, curiously enough, the pioneers of medieval times were very largely of priestly type, as witness the monk of Malmesbury.

The legends of the dawn of history, however, distribute the power of flight with less of prejudice. Egyptian sculpture gives the figure of winged men; the British Museum has made the winged Assyrian bulls familiar to many, and both the cuneiform records of Assyria and the hieroglyphs of Egypt record flights that in reality were never made. The desire fathered the story then, and until Clement Ader either hopped with his Avion, as is persisted by his critics, or flew, as is claimed by his friends.

While the origin of many legends is questionable, that of others is easy enough to trace, though not to prove. Among the credulous the significance of the name of a people of Asia Minor, the Capnobates, 'those who travel by smoke,' gave rise to the assertion that Montgolfier was not first in the field--or rather in the air--since surely this people must have been responsible for the first hot-air balloons. Far less questionable is the legend of Icarus, for here it is possible to trace a foundation of fact in the story. Such a tribe as Daedalus governed could have had hardly any knowledge of the rudiments of science, and even their ruler, seeing how easy it is for birds to sustain themselves in the air, might be excused for believing that he, if he fashioned wings for himself, could use them. In that belief, let it be assumed, Daedalus made his wings; the boy, Icarus, learning that his father had determined on an attempt at flight secured the wings and fastened them to his own shoulders. A cliff seemed the likeliest place for a 'take-off,' and Icarus leaped from the cliff edge only to find that the possession of wings was not enough to assure flight to a human being. The sea that to this day bears his name witnesses that he made the attempt and perished by it.

In this is assumed the bald story, from which might grow the legend of a wise king who ruled a peaceful people--'judged, sitting in the sun,' as Browning has it, and fashioned for himself wings with which he flew over the sea and where he would, until the prince, Icarus, desired to emulate him. Icarus, fastening the wings to his shoulders with wax, was so imprudent as to fly too near the sun, when the wax melted and he fell, to lie mourned of water-nymphs on the shores of waters thenceforth Icarian. Between what we have assumed to be the base of fact, and the legend which has been invested with such poetic grace in Greek story, there is no more than a century or so of re-telling might give to any event among a people so simple and yet so given to imagery.

We may set aside as pure fable the stories of the winged horse of Perseus, and the flights of Hermes as messenger of the gods. With them may be placed the story of Empedocles, who failed to take Etna seriously enough, and found himself caught by an eruption while within the crater, so that, flying to safety in some hurry, he left behind but one sandal to attest that he had sought refuge in space--in all probability, if he escaped at all, he flew, but not in the sense that the aeronaut understands it. But, bearing in mind the many men who tried to fly in historic times, the legend of Icarus and Daedalus, in spite of the impossible form in which it is presented, may rank with the story of the Saracen of Constantinople, or with that of Simon the Magician. A simple folk would naturally idealise the man and magnify his exploit, as they magnified the deeds of some strong man to make the legends of Hercules, and there, full-grown from a mere legend, is the first record of a pioneer of flying. Such a theory is not nearly so fantastic as that which makes the Capnobates, on the strength of their name, the inventors of hot-air balloons. However it may be, both in story and in picture, Icarus and his less conspicuous father have inspired the Caucasian mind, and the world is the richer for them.

Of the unsupported myths--unsupported, that is, by even a shadow of probability--there is no end. Although Latin legend approaches nearer to fact than the Greek in some cases, in others it shows a disregard for possibilities which renders it of far less account. Thus Diodorus of Sicily relates that one Abaris travelled round the world on an arrow of gold, and Cassiodorus and Glycas and their like told of mechanical birds that flew and sang and even laid eggs. More credible is the story of Aulus Gellius, who in his Attic Nights tells how Archytas, four centuries prior to the opening of the Christian era, made a wooden pigeon that actually flew by means of a mechanism of balancing weights and the breath of a mysterious spirit hidden within it. There may yet arise one credulous enough to state that the mysterious spirit was precursor of the internal combustion engine, but, however that may be, the pigeon of Archytas almost certainly existed, and perhaps it actually glided or flew for short distances--or else Aulus Gellius was an utter liar, like Cassiodorus and his fellows. In far later times a certain John Muller, better known as Regiomontanus, is stated to have made an artificial eagle which accompanied Charles V. on his entry to and exit from Nuremberg, flying above the royal procession. But, since Muller died in 1436 and Charles was born in 1500, Muller may be ruled out from among the pioneers of mechanical flight, and it may be concluded that the historian of this event got slightly mixed in his dates.

Thus far, we have but indicated how one may draw from the richest stores from which the Aryan mind draws inspiration, the Greek and Latin mythologies and poetic adaptations of history. The existing legends of flight, however, are not thus to be localised, for with two possible exceptions they belong to all the world and to every civilisation, however primitive. The two exceptions are the Aztec and the Chinese; regarding the first of these, the Spanish conquistadores destroyed such civilisation as existed in Tenochtitlan so thoroughly that, if legend of flight was among the Aztec records, it went with the rest; as to the Chinese, it is more than passing strange that they, who claim to have known and done everything while the first of history was shaping, even to antedating the discovery of gunpowder that was not made by Roger Bacon, have not yet set up a claim to successful handling of a monoplane some four thousand years ago, or at least to the patrol of the Gulf of Korea and the Mongolian frontier by a forerunner of the 'blimp.'

The Inca civilisation of Peru yields up a myth akin to that of Icarus, which tells how the chieftain Ayar Utso grew wings and visited the sun--it was from the sun, too, that the founders of the Peruvian Inca dynasty, Manco Capac and his wife Mama Huella Capac, flew to earth near Lake Titicaca, to make the only successful experiment in pure tyranny that the world has ever witnessed. Teutonic legend gives forth Wieland the Smith, who made himself a dress with wings and, clad in it, rose and descended against the wind and in spite of it. Indian mythology, in addition to the story of the demons and their rigid dirigible, already quoted, gives the story of Hanouam, who fitted himself with wings by means of which he sailed in the air and, according to his desire, landed in the sacred Lauka. Bladud, the ninth king of Britain, is said to have crowned his feats of wizardry by making himself wings and attempting to fly--but the effort cost him a broken neck. Bladud may have been as mythic as Uther, and again he may have been a very early pioneer. The Finnish epic, 'Kalevala,' tells how Ilmarinen the Smith 'forged an eagle of fire,' with 'boat's walls between the wings,' after which he 'sat down on the bird's back and bones,' and flew.

Pure myths, these, telling how the desire to fly was characteristic of every age and every people, and how, from time to time, there arose an experimenter bolder than his fellows, who made some attempt to translate desire into achievement. And the spirit that animated these pioneers, in a time when things new were accounted things accursed, for the most part, has found expression in this present century in the utter daring and disregard of both danger and pain that stamps the flying man, a type of humanity differing in spirit from his earthbound fellows as fully as the soldier differs from the priest.

Throughout mediaeval times, records attest that here and there some man believed in and attempted flight, and at the same time it is clear that such were regarded as in league with the powers of evil. There is the half-legend, half-history of Simon the Magician, who, in the third year of the reign of Nero announced that he would raise himself in the air, in order to assert his superiority over St Paul. The legend states that by the aid of certain demons whom he had prevailed on to assist him, he actually lifted himself in the air--but St Paul prayed him down again. He slipped through the claws of the demons and fell headlong on the Forum at Rome, breaking his neck. The 'demons' may have been some primitive form of hot-air balloon, or a glider with which the magician attempted to rise into the wind; more probably, however, Simon threatened to ascend and made the attempt with apparatus as unsuitable as Bladud's wings, paying the inevitable penalty. Another version of the story gives St Peter instead of St Paul as the one whose prayers foiled Simon--apart from the identity of the apostle, the two accounts are similar, and both define the attitude of the age toward investigation and experiment in things untried.

Another and later circumstantial story, with similar evidence of some fact behind it, is that of the Saracen of Constantinople, who, in the reign of the Emperor Comnenus--some little time before Norman William made Saxon Harold swear away his crown on the bones of the saints at Rouen--attempted to fly round the hippodrome at Constantinople, having Comnenus among the great throng who gathered to witness the feat. The Saracen chose for his starting-point a tower in the midst of the hippodrome, and on the top of the tower he stood, clad in a long white robe which was stiffened with rods so as to spread and catch the breeze, waiting for a favourable wind to strike on him. The wind was so long in coming that the spectators grew impatient. 'Fly, O Saracen!' they called to him. 'Do not keep us waiting so long while you try the wind!' Comnenus, who had present with him the Sultan of the Turks, gave it as his opinion that the experiment was both dangerous and vain, and, possibly in an attempt to controvert such statement, the Saracen leaned into the wind and 'rose like a bird 'at the outset. But the record of Cousin, who tells the story in his Histoire de Constantinople, states that 'the weight of his body having more power to drag him down than his artificial wings had to sustain him, he broke his bones, and his evil plight was such that he did not long survive.'

Obviously, the Saracen was anticipating Lilienthal and his gliders by some centuries; like Simon, a genuine experimenter--both legends bear the impress of fact supporting them. Contemporary with him, and belonging to the history rather than the legends of flight, was Oliver, the monk of Malmesbury, who in the year 1065 made himself wings after the pattern of those supposed to have been used by Daedalus, attaching them to his hands and feet and attempting to fly with them. Twysden, in his Historiae Anglicanae Scriptores X, sets forth the story of Oliver, who chose a high tower as his starting-point, and launched himself in the air. As a matter of course, he fell, permanently injuring himself, and died some time later.

After these, a gap of centuries, filled in by impossible stories of magical flight by witches, wizards, and the like--imagination was fertile in the dark ages, but the ban of the church was on all attempt at scientific development, especially in such a matter as the conquest of the air. Yet there were observers of nature who argued that since birds could raise themselves by flapping their wings, man had only to make suitable wings, flap them, and he too would fly. As early as the thirteenth century Roger Bacon, the scientific friar of unbounded inquisitiveness and not a little real genius, announced that there could be made 'some flying instrument, so that a man sitting in the middle and turning some mechanism may put in motion some artificial wings which may beat the air like a bird flying.' But being a cautious man, with a natural dislike for being burnt at the stake as a necromancer through having put forward such a dangerous theory, Roger added, 'not that I ever knew a man who had such an instrument, but I am particularly acquainted with the man who contrived one.' This might have been a lame defence if Roger had been brought to trial as addicted to black arts; he seems to have trusted to the inadmissibility of hearsay evidence.

Some four centuries later there was published a book entitled Perugia Augusta, written by one C. Crispolti of Perugia--the date of the work in question is 1648. In it is recorded that 'one day, towards the close of the fifteenth century, whilst many of the principal gentry had come to Perugia to honour the wedding of Giovanni Paolo Baglioni, and some lancers were riding down the street by his palace, Giovanni Baptisti Danti unexpectedly and by means of a contrivance of wings that he had constructed proportionate to the size of his body took off from the top of a tower near by, and with a horrible hissing sound flew successfully across the great Piazza, which was densely crowded. But (oh, horror of an unexpected accident!) he had scarcely flown three hundred paces on his way to a certain point when the mainstay of the left wing gave way, and, being unable to support himself with the right alone, he fell on a roof and was injured in consequence. Those who saw not only this flight, but also the wonderful construction of the framework of the wings, said--and tradition bears them out--that he several times flew over the waters of Lake Thrasimene to learn how he might gradually come to earth. But, notwithstanding his great genius, he never succeeded.'

This reads circumstantially enough, but it may be borne in mind that the date of writing is more than half a century later than the time of the alleged achievement--the story had had time to round itself out. Danti, however, is mentioned by a number of writers, one of whom states that the failure of his experiment was due to the prayers of some individual of a conservative turn of mind, who prayed so vigorously that Danti fell appropriately enough on a church and injured himself to such an extent as to put an end to his flying career. That Danti experimented, there is little doubt, in view of the volume of evidence on the point, but the darkness of the Middle Ages hides the real truth as to the results of his experiments. If he had actually flown over Thrasimene, as alleged, then in all probability both Napoleon and Wellington would have had air scouts at Waterloo.

Danti's story may be taken as fact or left as fable, and with it the period of legend or vague statement may be said to end--the rest is history, both of genuine experimenters and of charlatans. Such instances of legend as are given here are not a tithe of the whole, but there is sufficient in the actual history of flight to bar out more than this brief mention of the legends, which, on the whole, go farther to prove man's desire to fly than his study and endeavour to solve the problems of the air.

II. EARLY EXPERIMENTS

So far, the stories of the development of flight are either legendary or of more or less doubtful authenticity, even including that of Danti, who, although a man of remarkable attainments in more directions than that of attempted flight, suffers--so far as reputation is concerned--from the inexactitudes of his chroniclers; he may have soared over Thrasimene, as stated, or a mere hop with an ineffectual glider may have grown with the years to a legend of gliding flight. So far, too, there is no evidence of the study that the conquest of the air demanded; such men as made experiments either launched themselves in the air from some height with made-up wings or other apparatus, and paid the penalty, or else constructed some form of machine which would not leave the earth, and then gave up. Each man followed his own way, and there was no attempt--without the printing press and the dissemination of knowledge there was little possibility of attempt--on the part of any one to benefit by the failures of others.

Legend and doubtful history carries up to the fifteenth century, and then came

Leonardo da Vinci, first student of flight whose work endures to the present day. The world knows da Vinci as artist; his age knew him as architect, engineer, artist, and scientist in an age when science was a single study, comprising all knowledge from mathematics to medicine. He was, of course, in league with the devil, for in no other way could his range of knowledge and observation be explained by his contemporaries; he left a Treatise on the Flight of Birds in which are statements and deductions that had to be rediscovered when the Treatise had been forgotten--da Vinci anticipated modern knowledge as Plato anticipated modern thought, and blazed the first broad trail toward flight.

One Cuperus, who wrote a Treatise on the Excellence of Man, asserted that da Vinci translated his theories into practice, and actually flew, but the statement is unsupported. That he made models, especially on the helicopter principle, is past question; these were made of paper and wire, and actuated by springs of steel wire, which caused them to lift themselves in the air. It is, however, in the theories which he put forward that da Vinci's investigations are of greatest interest; these prove him a patient as well as a keen student of the principles of flight, and show that his manifold activities did not prevent him from devoting some lengthy periods to observations of bird flight.

'A bird,' he says in his Treatise, 'is an instrument working according to mathematical law, which instrument it is within the capacity of man to reproduce with all its movements, but not with a corresponding degree of strength, though it is deficient only in power of maintaining equilibrium. We may say, therefore, that such an instrument constructed by man is lacking in nothing except the life of the bird, and this life must needs be supplied from that of man. The life which resides in the bird's members will, without doubt, better conform to their needs than will that of a man which is separated from them, and especially in the almost imperceptible movements which produce equilibrium. But since we see that the bird is equipped for many apparent varieties of movement, we are able from this experience to deduce that the most rudimentary of these movements will be capable of being comprehended by man's understanding, and that he will to a great extent be able to provide against the destruction of that instrument of which he himself has become the living principle and the propeller.'

In this is the definite belief of da Vinci that man is capable of flight, together with a far more definite statement of the principles by which flight is to be achieved than any which had preceded it--and for that matter, than many that have succeeded it. Two further extracts from his work will show the exactness of his observations:--

'When a bird which is in equilibrium throws the centre of resistance of the wings behind the centre of gravity, then such a bird will descend with its head downward. This bird which finds itself in equilibrium shall have the centre of resistance of the wings more forward than the bird's centre of gravity; then such a bird will fall with its tail turned toward the earth.'

And again: 'A man, when flying, shall be free from the waist up, that he may be able to keep himself in equilibrium as he does in a boat, so that the centre of his gravity and of the instrument may set itself in equilibrium and change when necessity requires it to the changing of the centre of its resistance.'

Here, in this last quotation, are the first beginnings of the inherent stability which proved so great an advance in design, in this twentieth century. But the extracts given do not begin to exhaust the range of da Vinci's observations and deductions. With regard to bird flight, he observed that so long as a bird keeps its wings outspread it cannot fall directly to earth, but must glide down at an angle to alight--a small thing, now that the principle of the plane in opposition to the air is generally grasped, but da Vinci had to find it out. From observation he gathered how a bird checks its own speed by opposing tail and wing surface to the direction of flight, and thus alights at the proper 'landing speed.' He proved the existence of upward air currents by noting how a bird takes off from level earth with wings outstretched and motionless, and, in order to get an efficient substitute for the natural wing, he recommended that there be used something similar to the membrane of the wing of a bat--from this to the doped fabric of an aeroplane wing is but a small step, for both are equally impervious to air. Again, da Vinci recommended that experiments in flight be conducted at a good height from the ground, since, if equilibrium be lost through any cause, the height gives time to regain it. This recommendation, by the way, received ample support in the training areas of war pilots.

Man's muscles, said da Vinci, are fully sufficient to enable him to fly, for the larger birds, he noted, employ but a small part of their strength in keeping themselves afloat in the air--by this theory he attempted to encourage experiment, just as, when his time came, Borelli reached the opposite conclusion and discouraged it. That Borelli was right--so far--and da Vinci wrong, detracts not at all from the repute of the earlier investigator, who had but the resources of his age to support investigations conducted in the spirit of ages after.

His chief practical contributions to the science of flight--apart from numerous drawings which have still a value--are the helicopter or lifting screw, and the parachute. The former, as already noted, he made and proved effective in model form, and the principle which he demonstrated is that of the helicopter of to-day, on which sundry experimenters work spasmodically, in spite of the success of the plane with its driving propeller. As to the parachute, the idea was doubtless inspired by observation of the effect a bird produced by pressure of its wings against the direction of flight.

Da Vinci's conclusions, and his experiments, were forgotten easily by most of his contemporaries; his Treatise lay forgotten for nearly four centuries, overshadowed, mayhap, by his other work. There was, however, a certain Paolo Guidotti of Lucca, who lived in the latter half of the sixteenth century, and who attempted to carry da Vinci's theories--one of them, at least, into practice. For this Guidotti, who was by profession an artist and by inclination an investigator, made for himself wings, of which the framework was of whalebone; these he covered with feathers, and with them made a number of gliding flights, attaining considerable proficiency. He is said in the end to have made a flight of about four hundred yards, but this attempt at solving the problem ended on a house roof, where Guidotti broke his thigh bone. After that, apparently, he gave up the idea of flight, and went back to painting.

One other a Venetian architect named Veranzio, studied da Vinci's theory of the parachute, and found it correct, if contemporary records and even pictorial presentment are correct. Da Vinci showed his conception of a parachute as a sort of inverted square bag; Veranzio modified this to a 'sort of square sail extended by four rods of equal size and having four cords attached at the corners,' by means of which 'a man could without danger throw himself from the top of a tower or any high place. For though at the moment there may be no wind, yet the effort of his falling will carry up the wind, which the sail will hold, by which means he does not fall suddenly but descends little by little. The size of the sail should be measured to the man.' By this last, evidently, Veranzio intended to convey that the sheet must be of such content as would enclose sufficient air to support the weight of the parachutist.

Veranzio made his experiments about 1617-1618, but, naturally, they carried him no farther than the mere descent to earth, and since a descent is merely a descent, it is to be conjectured that he soon got tired of dropping from high roofs, and took to designing architecture instead of putting it to such a use. With the end of his experiments the work of da Vinci in relation to flying became neglected for nearly four centuries.

Apart from these two experimenters, there is little to record in the matter either of experiment or study until the seventeenth century. Francis Bacon, it is true, wrote about flying in his Sylva Sylvarum, and mentioned the subject in the New Atlantis, but, except for the insight that he showed even in superficial mention of any specific subject, he does not appear to have made attempt at serious investigation. 'Spreading of Feathers, thin and close and in great breadth will likewise bear up a great Weight,' says Francis, 'being even laid without Tilting upon the sides.' But a lesser genius could have told as much, even in that age, and though the great Sir Francis is sometimes adduced as one of the early students of the problems of flight, his writings will not sustain the reputation.

The seventeenth century, however, gives us three names, those of Borelli, Lana, and Robert Hooke, all of which take definite place in the history of flight. Borelli ranks as one of the great figures in the study of aeronautical problems, in spite of erroneous deductions through which he arrived at a purely negative conclusion with regard to the possibility of human flight.

Borelli was a versatile genius. Born in 1608, he was practically contemporary with Francesco Lana, and there is evidence that he either knew or was in correspondence with many prominent members of the Royal Society of Great Britain, more especially with John Collins, Dr Wallis, and Henry Oldenburgh, the then Secretary of the Society. He was author of a long list of scientific essays, two of which only are responsible for his fame, viz., Theorice Medicaearum Planetarum, published in Florence, and the better known posthumous De Motu Animalium. The first of these two is an astronomical study in which Borelli gives evidence of an instinctive knowledge of gravitation, though no definite expression is given of this. The second work, De Motu Animalium, deals with the mechanical action of the limbs of birds and animals and with a theory of the action of the internal organs. A section of the first part of this work, called De Volatu, is a study of bird flight; it is quite independent of Da Vinci's earlier work, which had been forgotten and remained unnoticed until near on the beginning of practical flight.

Marey, in his work, La Machine Animale, credits Borelli with the first correct idea of the mechanism of flight. He says: 'Therefore we must be allowed to render to the genius of Borelli the justice which is due to him, and only claim for ourselves the merit of having furnished the experimental demonstration of a truth already suspected.' In fact, all subsequent studies on this subject concur in making Borelli the first investigator who illustrated the purely mechanical theory of the action of a bird's wings.

Borelli's study is divided into a series of propositions in which he traces the principles of flight, and the mechanical actions of the wings of birds. The most interesting of these are the propositions in which he sets forth the method in which birds move their wings during flight and the manner in which the air offers resistance to the stroke of the wing. With regard to the first of these two points he says: 'When birds in repose rest on the earth their wings are folded up close against their flanks, but when wishing to start on their flight they first bend their legs and leap into the air. Whereupon the joints of their wings are straightened out to form a straight line at right angles to the lateral surface of the breast, so that the two wings, outstretched, are placed, as it were, like the arms of a cross to the body of the bird. Next, since the wings with their feathers attached form almost a plane surface, they are raised slightly above the horizontal, and with a most quick impulse beat down in a direction almost perpendicular to the wing-plane, upon the underlying air; and to so intense a beat the air, notwithstanding it to be fluid, offers resistance, partly by reason of its natural inertia, which seeks to retain it at rest, and partly because the particles of the air, compressed by the swiftness of the stroke, resist this compression by their elasticity, just like the hard ground. Hence the whole mass of the bird rebounds, making a fresh leap through the air; whence it follows that flight is simply a motion composed of successive leaps accomplished through the air. And I remark that a wing can easily beat the air in a direction almost perpendicular to its plane surface, although only a single one of the corners of the humerus bone is attached to the scapula, the whole extent of its base remaining free and loose, while the greater transverse feathers are joined to the lateral skin of the thorax. Nevertheless the wing can easily revolve about its base like unto a fan. Nor are there lacking tendon ligaments which restrain the feathers and prevent them from opening farther, in the same fashion that sheets hold in the sails of ships. No less admirable is nature's cunning in unfolding and folding the wings upwards, for she folds them not laterally, but by moving upwards edgewise the osseous parts wherein the roots of the feathers are inserted; for thus, without encountering the air's resistance the upward motion of the wing surface is made as with a sword, hence they can be uplifted with but small force. But thereafter when the wings are twisted by being drawn transversely and by the resistance of the air, they are flattened as has been declared and will be made manifest hereafter.'

Then with reference to the resistance to the air of the wings he explains: 'The air when struck offers resistance by its elastic virtue through which the particles of the air compressed by the wing-beat strive to expand again. Through these two causes of resistance the downward beat of the wing is not only opposed, but even caused to recoil with a reflex movement; and these two causes of resistance ever increase the more the down stroke of the wing is maintained and accelerated. On the other hand, the impulse of the wing is continuously diminished and weakened by the growing resistance. Hereby the force of the wing and the resistance become balanced; so that, manifestly, the air is beaten by the wing with the same force as the resistance to the stroke.'

He concerns himself also with the most difficult problem that confronts the flying man of to-day, namely, landing effectively, and his remarks on this subject would be instructive even to an air pilot of these days: 'Now the ways and means by which the speed is slackened at the end of a flight are these. The bird spreads its wings and tail so that their concave surfaces are perpendicular to the direction of motion; in this way, the spreading feathers, like a ship's sail, strike against the still air, check the speed, and so that most of the impetus may be stopped, the wings are flapped quickly and strongly forward, inducing a contrary motion, so that the bird absolutely or very nearly stops.'

At the end of his study Borelli came to a conclusion which militated greatly against experiment with any heavier-than-air apparatus, until well on into the nineteenth century, for having gone thoroughly into the subject of bird flight he states distinctly in his last proposition on the subject that 'It is impossible that men should be able to fly craftily by their own strength.' This statement, of course, remains true up to the present day for no man has yet devised the means by which he can raise himself in the air and maintain himself there by mere muscular effort.

From the time of Borelli up to the development of the steam engine it may be said that flight by means of any heavier-than-air apparatus was generally regarded as impossible, and apart from certain deductions which a little experiment would have shown to be doomed to failure, this method of flight was not followed up. It is not to be wondered at, when Borelli's exaggerated estimate of the strength expended by birds in proportion to their weight is borne in mind; he alleged that the motive force in birds' wings is 10,000 times greater than the resistance of their weight, and with regard to human flight he remarks:--

'When, therefore, it is asked whether men may be able to fly by their own strength, it must be seen whether the motive power of the pectoral muscles (the strength of which is indicated and measured by their size) is proportionately great, as it is evident that it must exceed the resistance of the weight of the whole human body 10,000 times, together with the weight of enormous wings which should be attached to the arms. And it is clear that the motive power of the pectoral muscles in men is much less than is necessary for flight, for in birds the bulk and weight of the muscles for flapping the wings are not less than a sixth part of the entire weight of the body. Therefore, it would be necessary that the pectoral muscles of a man should weigh more than a sixth part of the entire weight of his body; so also the arms, by flapping with the wings attached, should be able to exert a power 10,000 times greater than the weight of the human body itself. But they are far below such excess, for the aforesaid pectoral muscles do not equal a hundredth part of the entire weight of a man. Wherefore either the strength of the muscles ought to be increased or the weight of the human body must be decreased, so that the same proportion obtains in it as exists in birds. Hence it is deducted that the Icarian invention is entirely mythical because impossible, for it is not possible either to increase a man's pectoral muscles or to diminish the weight of the human body; and whatever apparatus is used, although it is possible to increase the momentum, the velocity or the power employed can never equal the resistance; and therefore wing flapping by the contraction of muscles cannot give out enough power to carry up the heavy body of a man.'

It may be said that practically all the conclusions which Borelli reached in his study were negative. Although contemporary with Lana, he perceived the one factor which rendered Lana's project for flight by means of vacuum globes an impossibility--he saw that no globe could be constructed sufficiently light for flight, and at the same time sufficiently strong to withstand the pressure of the outside atmosphere. He does not appear to have made any experiments in flying on his own account, having, as he asserts most definitely, no faith in any invention designed to lift man from the surface of the earth. But his work, from which only the foregoing short quotations can be given, is, nevertheless, of indisputable value, for he settled the mechanics of bird flight, and paved the way for those later investigators who had, first, the steam engine, and later the internal combustion engine--two factors in mechanical flight which would have seemed as impossible to Borelli as would wireless telegraphy to a student of Napoleonic times. On such foundations as his age afforded Borelli built solidly and well, so that he ranks as one of the greatest--if not actually the greatest--of the investigators into this subject before the age of steam.

The conclusion, that 'the motive force in birds' wings is apparently ten thousand times greater than the resistance of their weight,' is erroneous, of course, but study of the translation from which the foregoing excerpt is taken will show that the error detracts very little from the value of the work itself. Borelli sets out very definitely the mechanism of flight, in such fashion that he who runs may read. His reference to 'the use of a large vessel,' etc., concerns the suggestion made by Francesco Lana, who antedated Borelli's publication of De Motu Animalium by some ten years with his suggestion for an 'aerial ship,' as he called it. Lana's mind shows, as regards flight, a more imaginative twist; Borelli dived down into first causes, and reached mathematical conclusions; Lana conceived a theory and upheld it--theoretically, since the manner of his life precluded experiment.

Francesco Lana, son of a noble family, was born in 1631; in 1647 he was received as a novice into the Society of Jesus at Rome, and remained a pious member of the Jesuit society until the end of his life. He was greatly handicapped in his scientific investigations by the vows of poverty which the rules of the Order imposed on him. He was more scientist than priest all his life; for two years he held the post of Professor of Mathematics at Ferrara, and up to the time of his death, in 1687, he spent by far the greater part of his time in scientific research, He had the dubious advantage of living in an age when one man could cover the whole range of science, and this he seems to have done very thoroughly. There survives an immense work of his entitled, Magisterium Naturae et Artis, which embraces the whole field of scientific knowledge as that was developed in the period in which Lana lived. In an earlier work of his, published in Brescia in 1670, appears his famous treatise on the aerial ship, a problem which Lana worked out with thoroughness. He was unable to make practical experiments, and thus failed to perceive the one insuperable drawback to his project--of which more anon.

Only extracts from the translation of Lana's work can be given here, but sufficient can be given to show fully the means by which he designed to achieve the conquest of the air. He begins by mention of the celebrated pigeon of Archytas the Philosopher, and advances one or two theories with regard to the way in which this mechanical bird was constructed, and then he recites, apparently with full belief in it, the fable of Regiomontanus and the eagle that he is said to have constructed to accompany Charles V. on his entry into Nuremberg. In fact, Lana starts his work with a study of the pioneers of mechanical flying up to his own time, and then outlines his own devices for the construction of mechanical birds before proceeding to detail the construction of the aerial ship. Concerning primary experiments for this he says:--

'I will, first of all, presuppose that air has weight owing to the vapours and halations which ascend from the earth and seas to a height of many miles and surround the whole of our terraqueous globe; and this fact will not be denied by philosophers, even by those who may have but a superficial knowledge, because it can be proven by exhausting, if not all, at any rate the greater part of, the air contained in a glass vessel, which, if weighed before and after the air has been exhausted, will be found materially reduced in weight. Then I found out how much the air weighed in itself in the following manner. I procured a large vessel of glass, whose neck could be closed or opened by means of a tap, and holding it open I warmed it over a fire, so that the air inside it becoming rarified, the major part was forced out; then quickly shutting the tap to prevent the re-entry I weighed it; which done, I plunged its neck in water, resting the whole of the vessel on the surface of the water, then on opening the tap the water rose in the vessel and filled the greater part of it. I lifted the neck out of the water, released the water contained in the vessel, and measured and weighed its quantity and density, by which I inferred that a certain quantity of air had come out of the vessel equal in bulk to the quantity of water which had entered to refill the portion abandoned by the air. I again weighed the vessel, after I had first of all well dried it free of all moisture, and found it weighed one ounce more whilst it was full of air than when it was exhausted of the greater part, so that what it weighed more was a quantity of air equal in volume to the water which took its place. The water weighed 640 ounces, so I concluded that the weight of air compared with that of water was 1 to 640--that is to say, as the water which filled the vessel weighed 640 ounces, so the air which filled the same vessel weighed one ounce.'

Having thus detailed the method of exhausting air from a vessel, Lana goes on to assume that any large vessel can be entirely exhausted of nearly all the air contained therein. Then he takes Euclid's proposition to the effect that the superficial area of globes increases in the proportion of the square of the diameter, whilst the volume increases in the proportion of the cube of the same diameter, and he considers that if one only constructs the globe of thin metal, of sufficient size, and exhausts the air in the manner that he suggests, such a globe will be so far lighter than the surrounding atmosphere that it will not only rise, but will be capable of lifting weights. Here is Lana's own way of putting it:--

'But so that it may be enabled to raise heavier weights and to lift men in the air, let us take double the quantity of copper, 1,232 square feet, equal to 308 lbs. of copper; with this double quantity of copper we could construct a vessel of not only double the capacity, but of four times the capacity of the first, for the reason shown by my fourth supposition. Consequently the air contained in such a vessel will be 718 lbs. 4 2/3 ounces, so that if the air be drawn out of the vessel it will be 410 lbs. 4 2/3 ounces lighter than the same volume of air, and, consequently, will be enabled to lift three men, or at least two, should they weigh more than eight pesi each. It is thus manifest that the larger the ball or vessel is made, the thicker and more solid can the sheets of copper be made, because, although the weight will increase, the capacity of the vessel will increase to a greater extent and with it the weight of the air therein, so that it will always be capable to lift a heavier weight. From this it can be easily seen how it is possible to construct a machine which, fashioned like unto a ship, will float on the air.'

With four globes of these dimensions Lana proposed to make an aerial ship of the fashion shown in his quaint illustration. He is careful to point out a method by which the supporting globes for the aerial ship may be entirely emptied of air; (this is to be done by connecting to each globe a tube of copper which is 'at least a length of 47 modern Roman palm).' A small tap is to close this tube at the end nearest the globe, and then vessel and tube are to be filled with water, after which the tube is to be immersed in water and the tap opened, allowing the water to run out of the vessel, while no air enters. The tap is then closed before the lower end of the tube is removed from the water, leaving no air at all in the globe or sphere. Propulsion of this airship was to be accomplished by means of sails, and also by oars.

Lana antedated the modern propeller, and realised that the air would offer enough resistance to oars or paddle to impart motion to any vessel floating in it and propelled by these means, although he did not realise the amount of pressure on the air which would be necessary to accomplish propulsion. As a matter of fact, he foresaw and provided against practically all the difficulties that would be encountered in the working, as well as the making, of the aerial ship, finally coming up against what his religious training made an insuperable objection. This, again, is best told in his own words:--

'Other difficulties I do not foresee that could prevail against this invention, save one only, which to me seems the greatest of them all, and that is that God would surely never allow such a machine to be successful, since it would create many disturbances in the civil and political governments of mankind.'

He ends by saying that no city would be proof against surprise, while the aerial ship could set fire to vessels at sea, and destroy houses, fortresses, and cities by fire balls and bombs. In fact, at the end of his treatise on the subject, he furnishes a pretty complete resume of the activities of German Zeppelins.

As already noted, Lana himself, owing to his vows of poverty, was unable to do more than put his suggestions on paper, which he did with a thoroughness that has procured him a place among the really great pioneers of flying.

It was nearly 200 years before any attempt was made to realise his project; then, in 1843, M. Marey Monge set out to make the globes and the ship as Lana detailed them. Monge's experiments cost him the sum of 25,000 francs 75 centimes, which he expended purely from love of scientific investigation. He chose to make his globes of brass, about .004 in thickness, and weighing 1.465 lbs. to the square yard. Having made his sphere of this metal, he lined it with two thicknesses of tissue paper, varnished it with oil, and set to work to empty it of air. This, however, he never achieved, for such metal is incapable of sustaining the pressure of the outside air, as Lana, had he had the means to carry out experiments, would have ascertained. M. Monge's sphere could never be emptied of air sufficiently to rise from the earth; it ended in the melting-pot, ignominiously enough, and all that Monge got from his experiment was the value of the scrap metal and the satisfaction of knowing that Lana's theory could never be translated into practice.

Robert Hooke is less conspicuous than either Borelli or Lana; his work, which came into the middle of the seventeenth century, consisted of various experiments with regard to flight, from which emerged 'a Module, which by the help of Springs and Wings, raised and sustained itself in the air.' This must be reckoned as the first model flying machine which actually flew, except for da Vinci's helicopters; Hooke's model appears to have been of the flapping-wing type--he attempted to copy the motion of birds, but found from study and experiment that human muscles were not sufficient to the task of lifting the human body. For that reason, he says, 'I applied my mind to contrive a way to make artificial muscles,' but in this he was, as he expresses it, 'frustrated of my expectations.' Hooke's claim to fame rests mainly on his successful model; the rest of his work is of too scrappy a nature to rank as a serious contribution to the study of flight.

Contemporary with Hooke was one Allard, who, in France, undertook to emulate the Saracen of Constantinople to a certain extent. Allard was a tight-rope dancer who either did or was said to have done short gliding flights--the matter is open to question--and finally stated that he would, at St Germains, fly from the terrace in the king's presence. He made the attempt, but merely fell, as did the Saracen some centuries before, causing himself serious injury. Allard cannot be regarded as a contributor to the development of aeronautics in any way, and is only mentioned as typical of the way in which, up to the time of the Wright brothers, flying was regarded. Even unto this day there are many who still believe that, with a pair of wings, man ought to be able to fly, and that the mathematical data necessary to effective construction simply do not exist. This attitude was reasonable enough in an unlearned age, and Allard was one--a little more conspicuous than the majority--among many who made experiment in ignorance, with more or less danger to themselves and without practical result of any kind.

The seventeenth century was not to end, however, without practical experiment of a noteworthy kind in gliding flight. Among the recruits to the ranks of pioneers was a certain Besnier, a locksmith of Sable, who somewhere between 1675 and 1680 constructed a glider of which a crude picture has come down to modern times. The apparatus, as will be seen, consisted of two rods with hinged flaps, and the original designer of the picture seems to have had but a small space in which to draw, since obviously the flaps must have been much larger than those shown. Besnier placed the rods on his shoulders, and worked the flaps by cords attached to his hands and feet--the flaps opened as they fell, and closed as they rose, so the device as a whole must be regarded as a sort of flapping glider. Having by experiment proved his apparatus successful, Besnier promptly sold it to a travelling showman of the period, and forthwith set about constructing a second set, with which he made gliding flights of considerable height and distance. Like Lilienthal, Besnier projected himself into space from some height, and then, according to the contemporary records, he was able to cross a river of considerable size before coming to earth. It does not appear that he had any imitators, or that any advantage whatever was taken of his experiments; the age was one in which he would be regarded rather as a freak exhibitor than as a serious student, and possibly, considering his origin and the sale of his first apparatus to such a client, he regarded the matter himself as more in the nature of an amusement than as a discovery.

Borelli, coming at the end of the century, proved to his own satisfaction and that of his fellows that flapping wing flight was an impossibility; the capabilities of the plane were as yet undreamed, and the prime mover that should make the plane available for flight was deep in the womb of time. Da Vinci's work was forgotten--flight was an impossibility, or at best such a useless show as Besnier was able to give.

The eighteenth century was almost barren of experiment. Emanuel Swedenborg, having invented a new religion, set about inventing a flying machine, and succeeded theoretically, publishing the result of his investigations as follows:--

'Let a car or boat or some like object be made of light material such as cork or bark, with a room within it for the operator. Secondly, in front as well as behind, or all round, set a widely-stretched sail parallel to the machine forming within a hollow or bend which could be reefed like the sails of a ship. Thirdly, place wings on the sides, to be worked up and down by a spiral spring, these wings also to be hollow below in order to increase the force and velocity, take in the air, and make the resistance as great as may be required. These, too, should be of light material and of sufficient size; they should be in the shape of birds' wings, or the sails of a windmill, or some such shape, and should be tilted obliquely upwards, and made so as to collapse on the upward stroke and expand on the downward. Fourth, place a balance or beam below, hanging down perpendicularly for some distance with a small weight attached to its end, pendent exactly in line with the centre of gravity; the longer this beam is, the lighter must it be, for it must have the same proportion as the well-known vectis or steel-yard. This would serve to restore the balance of the machine if it should lean over to any of the four sides. Fifthly, the wings would perhaps have greater force, so as to increase the resistance and make the flight easier, if a hood or shield were placed over them, as is the case with certain insects. Sixthly, when the sails are expanded so as to occupy a great surface and much air, with a balance keeping them horizontal, only a small force would be needed to move the machine back and forth in a circle, and up and down. And, after it has gained momentum to move slowly upwards, a slight movement and an even bearing would keep it balanced in the air and would determine its direction at will.'

The only point in this worthy of any note is the first device for maintaining stability automatically--Swedenborg certainly scored a point there. For the rest, his theory was but theory, incapable of being put to practice--he does not appear to have made any attempt at advance beyond the mere suggestion.

Some ten years before his time the state of knowledge with regard to flying in Europe was demonstrated by an order granted by the King of Portugal to Friar Lourenzo de Guzman, who claimed to have invented a flying machine capable of actual flight. The order stated that 'In order to encourage the suppliant to apply himself with zeal toward the improvement of the new machine, which is capable of producing the effects mentioned by him, I grant unto him the first vacant place in my College of Barcelos or Santarem, and the first professorship of mathematics in my University of Coimbra, with the annual pension of 600,000 reis during his life.--Lisbon, 17th of March, 1709.'

What happened to Guzman when the non-existence of the machine was discovered is one of the things that is well outside the province of aeronautics. He was charlatan pure and simple, as far as actual flight was concerned, though he had some ideas respecting the design of hot-air balloons, according to Tissandier. (La Navigation Aerienne.) His flying machine was to contain, among other devices, bellows to produce artificial wind when the real article failed, and also magnets in globes to draw the vessel in an upward direction and maintain its buoyancy. Some draughtsman, apparently gifted with as vivid imagination as Guzman himself, has given to the world an illustration of the hypothetical vessel; it bears some resemblance to Lana's aerial ship, from which fact one draws obvious conclusions.

A rather amusing claim to solving the problem of flight was made in the middle of the eighteenth century by one Grimaldi, a 'famous and unique Engineer' who, as a matter of actual fact, spent twenty years in missionary work in India, and employed the spare time that missionary work left him in bringing his invention to a workable state. The invention is described as a 'box which with the aid of clockwork rises in the air, and goes with such lightness and strong rapidity that it succeeds in flying a journey of seven leagues in an hour. It is made in the fashion of a bird; the wings from end to end are 25 feet in extent. The body is composed of cork, artistically joined together and well fastened with metal wire, covered with parchment and feathers. The wings are made of catgut and whalebone, and covered also with the same parchment and feathers, and each wing is folded in three seams. In the body of the machine are contained thirty wheels of unique work, with two brass globes and little chains which alternately wind up a counterpoise; with the aid of six brass vases, full of a certain quantity of quicksilver, which run in some pulleys, the machine is kept by the artist in due equilibrium and balance. By means, then, of the friction between a steel wheel adequately tempered and a very heavy and surprising piece of lodestone, the whole is kept in a regulated forward movement, given, however, a right state of the winds, since the machine cannot fly so much in totally calm weather as in stormy. This prodigious machine is directed and guided by a tail seven palmi long, which is attached to the knees and ankles of the inventor by leather straps; by stretching out his legs, either to the right or to the left, he moves the machine in whichever direction he pleases.... The machine's flight lasts only three hours, after which the wings gradually close themselves, when the inventor, perceiving this, goes down gently, so as to get on his own feet, and then winds up the clockwork and gets himself ready again upon the wings for the continuation of a new flight. He himself told us that if by chance one of the wheels came off or if one of the wings broke, it is certain he would inevitably fall rapidly to the ground, and, therefore, he does not rise more than the height of a tree or two, as also he only once put himself in the risk of crossing the sea, and that was from Calais to Dover, and the same morning he arrived in London.'

And yet there are still quite a number of people who persist in stating that Bleriot was the first man to fly across the Channel!

A study of the development of the helicopter principle was published in France in 1868, when the great French engineer Paucton produced his Theorie de la Vis d'Archimede. For some inexplicable reason, Paucton was not satisfied with the term 'helicopter,' but preferred to call it a 'pterophore,' a name which, so far as can be ascertained, has not been adopted by any other writer or investigator. Paucton stated that, since a man is capable of sufficient force to overcome the weight of his own body, it is only necessary to give him a machine which acts on the air 'with all the force of which it is capable and at its utmost speed,' and he will then be able to lift himself in the air, just as by the exertion of all his strength he is able to lift himself in water. 'It would seem,' says Paucton, 'that in the pterophore, attached vertically to a carriage, the whole built lightly and carefully assembled, he has found something that will give him this result in all perfection. In construction, one would be careful that the machine produced the least friction possible, and naturally it ought to produce little, as it would not be at all complicated. The new Daedalus, sitting comfortably in his carriage, would by means of a crank give to the pterophore a suitable circular (or revolving) speed. This single pterophore would lift him vertically, but in order to move horizontally he should be supplied with a tail in the shape of another pterophore. When he wished to stop for a little time, valves fixed firmly across the end of the space between the blades would automatically close the openings through which the air flows, and change the pterophore into an unbroken surface which would resist the flow of air and retard the fall of the machine to a considerable degree.'

The doctrine thus set forth might appear plausible, but it is based on the common misconception that all the force which might be put into the helicopter or 'pterophore' would be utilised for lifting or propelling the vehicle through the air, just as a propeller uses all its power to drive a ship through water. But, in applying such a propelling force to the air, most of the force is utilised in maintaining aerodynamic support--as a matter of fact, more force is needed to maintain this support than the muscle of man could possibly furnish to a lifting screw, and even if the helicopter were applied to a full-sized, engine-driven air vehicle, the rate of ascent would depend on the amount of surplus power that could be carried. For example, an upward lift of 1,000 pounds from a propeller 15 feet in diameter would demand an expenditure of 50 horse-power under the best possible conditions, and in order to lift this load vertically through such atmospheric pressure as exists at sea-level or thereabouts, an additional 20 horsepower would be required to attain a rate of 11 feet per second--50 horse-power must be continually provided for the mere support of the load, and the additional 20 horse-power must be continually provided in order to lift it. Although, in model form, there is nothing quite so strikingly successful as the helicopter in the range of flying machines, yet the essential weight increases so disproportionately to the effective area that it is necessary to go but very little beyond model dimensions for the helicopter to become quite ineffective.

That is not to say that the lifting screw must be totally ruled out so far as the construction of aircraft is concerned. Much is still empirical, so far as this branch of aeronautics is concerned, and consideration of the structural features of a propeller goes to show that the relations of essential weight and effective area do not altogether apply in practice as they stand in theory. Paucton's dream, in some modified form, may yet become reality--it is only so short a time ago as 1896 that Lord Kelvin stated he had not the smallest molecule of faith in aerial navigation, and since the whole history of flight consists in proving the impossible possible, the helicopter may yet challenge the propelled plane surface for aerial supremacy.

It does not appear that Paucton went beyond theory, nor is there in his theory any advance toward practical flight--da Vinci could have told him as much as he knew. He was followed by Meerwein, who invented an apparatus apparently something between a flapping wing machine and a glider, consisting of two wings, which were to be operated by means of a rod; the venturesome one who would fly by means of this apparatus had to lie in a horizontal position beneath the wings to work the rod. Meerwein deserves a place of mention, however, by reason of his investigations into the amount of surface necessary to support a given weight. Taking that weight at 200 pounds--which would allow for the weight of a man and a very light apparatus--he estimated that 126 square feet would be necessary for support. His pamphlet, published at Basle in 1784, shows him to have been a painstaking student of the potentialities of flight.

Jean-Pierre Blanchard, later to acquire fame in connection with balloon flight, conceived and described a curious vehicle, of which he even announced trials as impending. His trials were postponed time after time, and it appears that he became convinced in the end of the futility of his device, being assisted to such a conclusion by Lalande, the astronomer, who repeated Borelli's statement that it was impossible for man ever to fly by his own strength. This was in the closing days of the French monarchy, and the ascent of the Montgolfiers' first hot-air balloon in 1783--which shall be told more fully in its place--put an end to all French experiments with heavier-than-air apparatus, though in England the genius of Cayley was about to bud, and even in France there were those who understood that ballooning was not true flight.

III. SIR GEORGE CAYLEY--THOMAS WALKER

On the fifth of June, 1783, the Montgolfiers' hot-air balloon rose at Versailles, and in its rising divided the study of the conquest of the air into two definite parts, the one being concerned with the propulsion of gas lifted, lighter-than-air vehicles, and the other being crystallised in one sentence by Sir George Cayley: 'The whole problem,' he stated, 'is confined within these limits, viz.: to make a surface support a given weight by the application of power to the resistance of the air.' For about ten years the balloon held the field entirely, being regarded as the only solution of the problem of flight that man could ever compass. So definite for a time was this view on the eastern side of the Channel that for some years practically all the progress that was made in the development of power-driven planes was made in Britain.

In 1800 a certain Dr Thomas Young demonstrated that certain curved surfaces suspended by a thread moved into and not away from a horizontal current of air, but the demonstration, which approaches perilously near to perpetual motion if the current be truly horizontal, has never been successfully repeated, so that there is more than a suspicion that Young's air-current was NOT horizontal. Others had made and were making experiments on the resistance offered to the air by flat surfaces, when Cayley came to study and record, earning such a place among the pioneers as to win the title of 'father of British aeronautics.'

Cayley was a man in advance of his time, in many ways. Of independent means, he made the grand tour which was considered necessary to the education of every young man of position, and during this excursion he was more engaged in studies of a semi-scientific character than in the pursuits that normally filled such a period. His various writings prove that throughout his life aeronautics was the foremost subject in his mind; the Mechanic's Magazine, Nicholson's Journal, the Philosophical Magazine, and other periodicals of like nature bear witness to Cayley's continued research into the subject of flight. He approached the subject after the manner of the trained scientist, analysing the mechanical properties of air under chemical and physical action. Then he set to work to ascertain the power necessary for aerial flight, and was one of the first to enunciate the fallacy of the hopes of successful flight by means of the steam engine of those days, owing to the fact that it was impossible to obtain a given power with a given weight.

Yet his conclusions on this point were not altogether negative, for as early as 1810 he stated that he could construct a balloon which could travel with passengers at 20 miles an hour--he was one of the first to consider the possibilities of applying power to a balloon. Nearly thirty years later--in 1837--he made the first attempt at establishing an aeronautical society, but at that time the power-driven plane was regarded by the great majority as an absurd dream of more or less mad inventors, while ballooning ranked on about the same level as tight-rope walking, being considered an adjunct to fairs and fetes, more a pastime than a study.

Up to the time of his death, in 1857, Cayley maintained his study of aeronautical matters, and there is no doubt whatever that his work went far in assisting the solution of the problem of air conquest. His principal published work, a monograph entitled Aerial Navigation, has been republished in the admirable series of 'Aeronautical Classics' issued by the Royal Aeronautical Society. He began this work by pointing out the impossibility of flying by means of attached wings, an impossibility due to the fact that, while the pectoral muscles of a bird account for more than two-thirds of its whole muscular strength, in a man the muscles available for flying, no matter what mechanism might be used, would not exceed one-tenth of his total strength.

Cayley did not actually deny the possibility of a man flying by muscular effort, however, but stated that 'the flight of a strong man by great muscular exertion, though a curious and interesting circumstance, inasmuch as it will probably be the means of ascertaining finis power and supplying the basis whereon to improve it, would be of little use.'

From this he goes on to the possibility of using a Boulton and Watt steam engine to develop the power necessary for flight, and in this he saw a possibility of practical result. It is worthy of note that in this connection he made mention of the forerunner of the modern internal combustion engine; 'The French,' he said, 'have lately shown the great power produced by igniting inflammable powders in closed vessels, and several years ago an engine was made to work in this country in a similar manner by inflammation of spirit of tar.' In a subsequent paragraph of his monograph he anticipates almost exactly the construction of the Lenoir gas engine, which came into being more than fifty-five years after his monograph was published.

Certain experiments detailed in his work were made to ascertain the size of the surface necessary for the support of any given weight. He accepted a truism of to-day in pointing out that in any matters connected with aerial investigation, theory and practice are as widely apart as the poles. Inclined at first to favour the helicopter principle, he finally rejected this in favour of the plane, with which he made numerous experiments. During these, he ascertained the peculiar advantages of curved surfaces, and saw the necessity of providing both vertical and horizontal rudders in order to admit of side steering as well as the control of ascent and descent, and for preserving equilibrium. He may be said to have anticipated the work of Lilienthal and Pilcher, since he constructed and experimented with a fixed surface glider. 'It was beautiful,' he wrote concerning this, 'to see this noble white bird sailing majestically from the top of a hill to any given point of the plain below it with perfect steadiness and safety, according to the set of its rudder, merely by its own weight, descending at an angle of about eight degrees with the horizon.'

It is said that he once persuaded his gardener to trust himself in this glider for a flight, but if Cayley himself ventured a flight in it he has left no record of the fact. The following extract from his work, Aerial Navigation, affords an instance of the thoroughness of his investigations, and the concluding paragraph also shows his faith in the ultimate triumph of mankind in the matter of aerial flight:--

'The act of flying requires less exertion than from the appearance is supposed. Not having sufficient data to ascertain the exact degree of propelling power exerted by birds in the act of flying, it is uncertain what degree of energy may be required in this respect for vessels of aerial navigation; yet when we consider the many hundreds of miles of continued flight exerted by birds of passage, the idea of its being only a small effort is greatly corroborated. To apply the power of the first mover to the greatest advantage in producing this effect is a very material point. The mode universally adopted by Nature is the oblique waft of the wing. We have only to choose between the direct beat overtaking the velocity of the current, like the oar of a boat, or one applied like the wing, in some assigned degree of obliquity to it. Suppose 35 feet per second to be the velocity of an aerial vehicle, the oar must be moved with this speed previous to its being able to receive any resistance; then if it be only required to obtain a pressure of one-tenth of a lb. upon each square foot it must exceed the velocity of the current 7.3 feet per second. Hence its whole velocity must be 42.5 feet per second. Should the same surface be wafted downward like a wing with the hinder edge inclined upward in an angle of about 50 deg. 40 feet to the current it will overtake it at a velocity of 3.5 feet per second; and as a slight unknown angle of resistance generates a lb. pressure per square foot at this velocity, probably a waft of a little more than 4 feet per second would produce this effect, one-tenth part of which would be the propelling power. The advantage of this mode of application compared with the former is rather more than ten to one.

'In continuing the general principles of aerial navigation, for the practice of the art, many mechanical difficulties present themselves which require a considerable course of skilfully applied experiments before they can be overcome; but, to a certain extent, the air has already been made navigable, and no one who has seen the steadiness with which weights to the amount of ten stone (including four stone, the weight of the machine) hover in the air can doubt of the ultimate accomplishment of this object.'

This extract from his work gives but a faint idea of the amount of research for which Cayley was responsible. He had the humility of the true investigator in scientific problems, and so far as can be seen was never guilty of the great fault of so many investigators in this subject--that of making claims which he could not support. He was content to do, and pass after having recorded his part, and although nearly half a century had to pass between the time of his death and the first actual flight by means of power-driven planes, yet he may be said to have contributed very largely to the solution of the problem, and his name will always rank high in the roll of the pioneers of flight.

Practically contemporary with Cayley was Thomas Walker, concerning whom little is known save that he was a portrait painter of Hull, where was published his pamphlet on The Art of Flying in 1810, a second and amplified edition being produced, also in Hull, in 1831. The pamphlet, which has been reproduced in extenso in the Aeronautical Classics series published by the Royal Aeronautical Society, displays a curious mixture of the true scientific spirit and colossal conceit. Walker appears to have been a man inclined to jump to conclusions, which carried him up to the edge of discovery and left him vacillating there.

The study of the two editions of his pamphlet side by side shows that their author made considerable advances in the practicability of his designs in the 21 intervening years, though the drawings which accompany the text in both editions fail to show anything really capable of flight. The great point about Walker's work as a whole is its suggestiveness; he did not hesitate to state that the 'art' of flying is as truly mechanical as that of rowing a boat, and he had some conception of the necessary mechanism, together with an absolute conviction that he knew all there was to be known. 'Encouraged by the public,' he says, 'I would not abandon my purpose of making still further exertions to advance and complete an art, the discovery of the TRUE PRINCIPLES (the italics are Walker's own) of which, I trust, I can with certainty affirm to be my own.'

The pamphlet begins with Walker's admiration of the mechanism of flight as displayed by birds. 'It is now almost twenty years,' he says, 'since I was first led to think, by the study of birds and their means of flying, that if an artificial machine were formed with wings in exact imitation of the mechanism of one of those beautiful living machines, and applied in the very same way upon the air, there could be no doubt of its being made to fly, for it is an axiom in philosophy that the same cause will ever produce the same effect.' With this he confesses his inability to produce the said effect through lack of funds, though he clothes this delicately in the phrase 'professional avocations and other circumstances.' Owing to this inability he published his designs that others might take advantage of them, prefacing his own researches with a list of the very early pioneers, and giving special mention to Friar Bacon, Bishop Wilkins, and the Portuguese friar, De Guzman. But, although he seems to suggest that others should avail themselves of his theoretical knowledge, there is a curious incompleteness about the designs accompanying his work, and about the work itself, which seems to suggest that he had more knowledge to impart than he chose to make public--or else that he came very near to complete solution of the problem of flight, and stayed on the threshold without knowing it.

After a dissertation upon the history and strength of the condor, and on the differences between the weights of birds, he says: 'The following observations upon the wonderful difference in the weight of some birds, with their apparent means of supporting it in their flight, may tend to remove some prejudices against my plan from the minds of some of my readers. The weight of the humming-bird is one drachm, that of the condor not less than four stone. Now, if we reduce four stone into drachms we shall find the condor is 14,336 times as heavy as the humming-bird. What an amazing disproportion of weight! Yet by the same mechanical use of its wings the condor can overcome the specific gravity of its body with as much ease as the little humming-bird. But this is not all. We are informed that this enormous bird possesses a power in its wings, so far exceeding what is necessary for its own conveyance through the air, that it can take up and fly away with a whole sheer in its talons, with as much ease as an eagle would carry off, in the same manner, a hare or a rabbit. This we may readily give credit to, from the known fact of our little kestrel and the sparrow-hawk frequently flying off with a partridge, which is nearly three times the weight of these rapacious little birds.'

After a few more observations he arrives at the following conclusion: 'By attending to the progressive increase in the weight of birds, from the delicate little humming-bird up to the huge condor, we clearly discover that the addition of a few ounces, pounds, or stones, is no obstacle to the art of flying; the specific weight of birds avails nothing, for by their possessing wings large enough, and sufficient power to work them, they can accomplish the means of flying equally well upon all the various scales and dimensions which we see in nature. Such being a fact, in the name of reason and philosophy why shall not man, with a pair of artificial wings, large enough, and with sufficient power to strike them upon the air, be able to produce the same effect?'

Walker asserted definitely and with good ground that muscular effort applied without mechanism is insufficient for human flight, but he states that if an aeronautical boat were constructed so that a man could sit in it in the same manner as when rowing, such a man would be able to bring into play his whole bodily strength for the purpose of flight, and at the same time would be able to get an additional advantage by exerting his strength upon a lever. At first he concluded there must be expansion of wings large enough to resist in a sufficient degree the specific gravity of whatever is attached to them, but in the second edition of his work he altered this to 'expansion of flat passive surfaces large enough to reduce the force of gravity so as to float the machine upon the air with the man in it.' The second requisite is strength enough to strike the wings with sufficient force to complete the buoyancy and give a projectile motion to the machine. Given these two requisites, Walker states definitely that flying must be accomplished simply by muscular exertion. 'If we are secure of these two requisites, and I am very confident we are, we may calculate upon the success of flight with as much certainty as upon our walking.'

Walker appears to have gained some confidence from the experiments of a certain M. Degen, a watchmaker of Vienna, who, according to the Monthly Magazine of September, 1809, invented a machine by means of which a person might raise himself into the air. The said machine, according to the magazine, was formed of two parachutes which might be folded up or extended at pleasure, while the person who worked them was placed in the centre. This account, however, was rather misleading, for the magazine carefully avoided mention of a balloon to which the inventor fixed his wings or parachutes. Walker, knowing nothing of the balloon, concluded that Degen actually raised himself in the air, though he is doubtful of the assertion that Degen managed to fly in various directions, especially against the wind.

Walker, after considering Degen and all his works, proceeds to detail his own directions for the construction of a flying machine, these being as follows: 'Make a car of as light material as possible, but with sufficient strength to support a man in it; provide a pair of wings about four feet each in length; let them be horizontally expanded and fastened upon the top edge of each side of the car, with two joints each, so as to admit of a vertical motion to the wings, which motion may be effected by a man sitting and working an upright lever in the middle of the car. Extend in the front of the car a flat surface of silk, which must be stretched out and kept fixed in a passive state; there must be the same fixed behind the car; these two surfaces must be perfectly equal in length and breadth and large enough to cover a sufficient quantity of air to support the whole weight as nearly in equilibrium as possible, thus we shall have a great sustaining power in those passive surfaces and the active wings will propel the car forward.'

A description of how to launch this car is subsequently given: 'It becomes necessary,' says the theorist, 'that I should give directions how it may be launched upon the air, which may be done by various means; perhaps the following method may be found to answer as well as any: Fix a poll upright in the earth, about twenty feet in height, with two open collars to admit another poll to slide upwards through them; let there be a sliding platform made fast upon the top of the sliding poll; place the car with a man in it upon the platform, then raise the platform to the height of about thirty feet by means of the sliding poll, let the sliding poll and platform suddenly fall down, the car will then be left upon the air, and by its pressing the air a projectile force will instantly propel the car forward; the man in the car must then strike the active wings briskly upon the air, which will so increase the projectile force as to become superior to the force of gravitation, and if he inclines his weight a little backward, the projectile impulse will drive the car forward in an ascending direction. When the car is brought to a sufficient altitude to clear the tops of hills, trees, buildings, etc., the man, by sitting a little forward on his seat, will then bring the wings upon a horizontal plane, and by continuing the action of the wings he will be impelled forward in that direction. To descend, he must desist from striking the wings, and hold them on a level with their joints; the car will then gradually come down, and when it is within five or six feet of the ground the man must instantly strike the wings downwards, and sit as far back as he can; he will by this means check the projectile force, and cause the car to alight very gently with a retrograde motion. The car, when up in the air, may be made to turn to the right or to the left by forcing out one of the fins, having one about eighteen inches long placed vertically on each side of the car for that purpose, or perhaps merely by the man inclining the weight of his body to one side.'

Having stated how the thing is to be done, Walker is careful to explain that when it is done there will be in it some practical use, notably in respect of the conveyance of mails and newspapers, or the saving of life at sea, or for exploration, etc. It might even reduce the number of horses kept by man for his use, by means of which a large amount of land might be set free for the growth of food for human consumption.

At the end of his work Walker admits the idea of steam power for driving a flying machine in place of simple human exertion, but he, like Cayley, saw a drawback to this in the weight of the necessary engine. On the whole, he concluded, navigation of the air by means of engine power would be mostly confined to the construction of navigable balloons.

As already noted, Walker's work is not over practical, and the foregoing extract includes the most practical part of it; the rest is a series of dissertations on bird flight, in which, evidently, the portrait painter's observations were far less thorough than those of da Vinci or Borelli. Taken on the whole, Walker was a man with a hobby; he devoted to it much time and thought, but it remained a hobby, nevertheless. His observations have proved useful enough to give him a place among the early students of flight, but a great drawback to his work is the lack of practical experiment, by means of which alone real advance could be made; for, as Cayley admitted, theory and practice are very widely separated in the study of aviation, and the whole history of flight is a matter of unexpected results arising from scarcely foreseen causes, together with experiment as patient as daring.

IV. THE MIDDLE NINETEENTH CENTURY

Both Cayley and Walker were theorists, though Cayley supported his theoretical work with enough of practice to show that he studied along right lines; a little after his time there came practical men who brought to being the first machine which actually flew by the application of power. Before their time, however, mention must be made of the work of George Pocock of Bristol, who, somewhere about 1840 invented what was described as a 'kite carriage,' a vehicle which carried a number of persons, and obtained its motive power from a large kite. It is on record that, in the year 1846 one of these carriages conveyed sixteen people from Bristol to London. Another device of Pocock's was what he called a 'buoyant sail,' which was in effect a man-lifting kite, and by means of which a passenger was actually raised 100 yards from the ground, while the inventor's son scaled a cliff 200 feet in height by means of one of these, 'buoyant sails.' This constitutes the first definitely recorded experiment in the use of man-lifting kites. A History of the Charvolant or Kite-carriage, published in London in 1851, states that 'an experiment of a bold and very novel character was made upon an extensive down, where a large wagon with a considerable load was drawn along, whilst this huge machine at the same time carried an observer aloft in the air, realising almost the romance of flying.'

Experimenting, two years after the appearance of the 'kite-carriage,' on the helicopter principle, W. H. Phillips constructed a model machine which weighed two pounds; this was fitted with revolving fans, driven by the combustion of charcoal, nitre, and gypsum, producing steam which, discharging into the air, caused the fans to revolve. The inventor stated that 'all being arranged, the steam was up in a few seconds, when the whole apparatus spun around like any top, and mounted into the air faster than a bird; to what height it ascended I had no means of ascertaining; the distance travelled was across two fields, where, after a long search, I found the machine minus the wings, which had been torn off in contact with the ground.' This could hardly be described as successful flight, but it was an advance in the construction of machines on the helicopter principle, and it was the first steam-driven model of the type which actually flew. The invention, however, was not followed up.

After Phillips, we come to the great figures of the middle nineteenth century, W. S. Henson and John Stringfellow. Cayley had shown, in 1809, how success might be attained by developing the idea of the plane surface so driven as to take advantage of the resistance offered by the air, and Henson, who as early as 1840 was experimenting with model gliders and light steam engines, evolved and patented an idea for something very nearly resembling the monoplane of the early twentieth century. His patent, No. 9478, of the year 1842 explains the principle of the machine as follows:--

In order that the description hereafter given be rendered clear, I will first shortly explain the principle on which the machine is constructed. If any light and flat or nearly flat article be projected or thrown edgewise in a slightly inclined position, the same will rise on the air till the force exerted is expended, when the article so thrown or projected will descend; and it will readily be conceived that, if the article so projected or thrown possessed in itself a continuous power or force equal to that used in throwing or projecting it, the article would continue to ascend so long as the forward part of the surface was upwards in respect to the hinder part, and that such article, when the power was stopped, or when the inclination was reversed, would descend by gravity aided by the force of the power contained in the article, if the power be continued, thus imitating the flight of a bird.

Now, the first part of my invention consists of an apparatus so constructed as to offer a very extended surface or plane of a light yet strong construction, which will have the same relation to the general machine which the extended wings of a bird have to the body when a bird is skimming in the air; but in place of the movement or power for onward progress being obtained by movement of the extended surface or plane, as is the case with the wings of birds, I apply suitable paddle-wheels or other proper mechanical propellers worked by a steam or other sufficiently light engine, and thus obtain the requisite power for onward movement to the plane or extended surface; and in order to give control as to the upward and downward direction of such a machine I apply a tail to the extended surface which is capable of being inclined or raised, so that when the power is acting to propel the machine, by inclining the tail upwards, the resistance offered by the air will cause the machine to rise on the air; and, on the contrary, when the inclination of the tail is reversed, the machine will immediately be propelled downwards, and pass through a plane more or less inclined to the horizon as the inclination of the tail is greater or less; and in order to guide the machine as to the lateral direction which it shall take, I apply a vertical rudder or second tail, and, according as the same is inclined in one direction or the other, so will be the direction of the machine.'

The machine in question was very large, and differed very little from the modern monoplane; the materials were to be spars of bamboo and hollow wood, with diagonal wire bracing. The surface of the planes was to amount to 4,500 square feet, and the tail, triangular in form (here modern practice diverges) was to be 1,500 square feet. The inventor estimated that there would be a sustaining power of half a pound per square foot, and the driving power was to be supplied by a steam engine of 25 to 30 horse-power, driving two six-bladed propellers. Henson was largely dependent on Stringfellow for many details of his design, more especially with regard to the construction of the engine.

The publication of the patent attracted a great amount of public attention, and the illustrations in contemporary journals, representing the machine flying over the pyramids and the Channel, anticipated fact by sixty years and more; the scientific world was divided, as it was up to the actual accomplishment of flight, as to the value of the invention.

Strongfellow and Henson became associated after the conception of their design, with an attorney named Colombine, and a Mr Marriott, and between the four of them a project grew for putting the whole thing on a commercial basis--Henson and Stringfellow were to supply the idea; Marriott, knowing a member of Parliament, would be useful in getting a company incorporated, and Colombine would look after the purely legal side of the business. Thus an application was made by Mr Roebuck, Marriott's M.P., for an act of incorporation for 'The Aerial Steam Transit Company,' Roebuck moving to bring in the bill on the 24th of March, 1843. The prospectus, calling for funds for the development of the invention, makes interesting reading at this stage of aeronautical development; it was as follows:

PROPOSAL.

For subscriptions of sums of L100, in furtherance of an Extraordinary Invention not at present safe to be developed by securing the necessary Patents, for which three times the sum advanced, namely, L300, is conditionally guaranteed for each subscription on February 1, 1844, in case of the anticipations being realised, with the option of the subscribers being shareholders for the large amount if so desired, but not otherwise.

---------An Invention has recently been discovered, which if ultimately successful will be without parallel even in the age which introduced to the world the wonderful effects of gas and of steam.

The discovery is of that peculiar nature, so simple in principle yet so perfect in all the ingredients required for complete and permanent success, that to promulgate it at present would wholly defeat its development by the immense competition which would ensue, and the views of the originator be entirely frustrated.

This work, the result of years of labour and study, presents a wonderful instance of the adaptation of laws long since proved to the scientific world combined with established principles so judiciously and carefully arranged, as to produce a discovery perfect in all its parts and alike in harmony with the laws of Nature and of science.

The Invention has been subjected to several tests and examinations and the results are most satisfactory so much so that nothing but the completion of the undertaking is required to determine its practical operation, which being once established its utility is undoubted, as it would be a necessary possession of every empire, and it were hardly too much to say, of every individual of competent means in the civilised world.

Its qualities and capabilities are so vast that it were impossible and, even if possible, unsafe to develop them further, but some idea may be formed from the fact that as a preliminary measure patents in Great Britain Ireland, Scotland, the Colonies, France, Belgium, and the United States, and every other country where protection to the first discoveries of an Invention is granted, will of necessity be immediately obtained, and by the time these are perfected, which it is estimated will be in the month of February, the Invention will be fit for Public Trial, but until the Patents are sealed any further disclosure would be most dangerous to the principle on which it is based.

Under these circumstances, it is proposed to raise an immediate sum of L2,000 in furtherance of the Projector's views, and as some protection to the parties who may embark in the matter, that this is not a visionary plan for objects imperfectly considered, Mr Colombine, to whom the secret has been confided, has allowed his name to be used on the occasion, and who will if referred to corroborate this statement, and convince any inquirer of the reasonable prospects of large pecuniary results following the development of the Invention.

It is, therefore, intended to raise the sum of L2,000 in twenty sums of L100 each (of which any subscriber may take one or more not exceeding five in number to be held by any individual) the amount of which is to be paid into the hands of Mr Colombine as General Manager of the concern to be by him appropriated in procuring the several Patents and providing the expenses incidental to the works in progress. For each of which sums of L100 it is intended and agreed that twelve months after the 1st February next, the several parties subscribing shall receive as an equivalent for the risk to be run the sum of L300 for each of the sums of L100 now subscribed, provided when the time arrives the Patents shall be found to answer the purposes intended.

As full and complete success is alone looked to, no moderate or imperfect benefit is to be anticipated, but the work, if it once passes the necessary ordeal, to which inventions of every kind must be first subject, will then be regarded by every one as the most astonishing discovery of modern times; no half success can follow, and therefore the full nature of the risk is immediately ascertained.

The intention is to work and prove the Patent by collective instead of individual aid as less hazardous at first end more advantageous in the result for the Inventor, as well as others, by having the interest of several engaged in aiding one common object--the development of a Great Plan. The failure is not feared, yet as perfect success might, by possibility, not ensue, it is necessary to provide for that result, and the parties concerned make it a condition that no return of the subscribed money shall be required, if the Patents shall by any unforeseen circumstances not be capable of being worked at all; against which, the first application of the money subscribed, that of securing the Patents, affords a reasonable security, as no one without solid grounds would think of such an expenditure.

It is perfectly needless to state that no risk or responsibility of any kind can arise beyond the payment of the sum to be subscribed under any circumstances whatever.

As soon as the Patents shall be perfected and proved it is contemplated, so far as may be found practicable, to further the great object in view a Company shall be formed but respecting which it is unnecessary to state further details, than that a preference will be given to all those persons who now subscribe, and to whom shares shall be appropriated according to the larger amount (being three times the sum to be paid by each person) contemplated to be returned as soon as the success of the Invention shall have been established, at their option, or the money paid, whereby the Subscriber will have the means of either withdrawing with a large pecuniary benefit, or by continuing his interest in the concern lay the foundation for participating in the immense benefit which must follow the success of the plan.

It is not pretended to conceal that the project is a speculation--all parties believe that perfect success, and thence incalculable advantage of every kind, will follow to every individual joining in this great undertaking; but the Gentlemen engaged in it wish that no concealment of the consequences, perfect success, or possible failure, should in the slightest degree be inferred. They believe this will prove the germ of a mighty work, and in that belief call for the operation of others with no visionary object, but a legitimate one before them, to attain that point where perfect success will be secured from their combined exertions.

All applications to be made to D. E. Colombine, Esquire, 8 Carlton Chambers, Regent Street.

The applications did not materialise, as was only to be expected in view of the vagueness of the proposals. Colombine did some advertising, and Mr Roebuck expressed himself as unwilling to proceed further in the venture. Henson experimented with models to a certain extent, while Stringfellow looked for funds for the construction of a full-sized monoplane. In November of 1843 he suggested that he and Henson should construct a large model out of their own funds. On Henson's suggestion Colombine and Marriott were bought out as regards the original patent, and Stringfellow and Henson entered into an agreement and set to work.

Their work is briefly described in a little pamphlet by F. J. Stringfellow, entitled *A few Remarks on what has been done with screw-propelled Aero-plane Machines from 1809 to 1892*. The author writes with regard to the work that his father and Henson undertook:--

'They commenced the construction of a small model operated by a spring, and laid down the larger model 20 ft. from tip to tip of planes, 3 1/2 ft. wide, giving 70 ft. of sustaining surface, about 10 more in the tail. The making of this model required great consideration; various supports for the wings were tried, so as to combine lightness with firmness, strength and rigidity.

'The planes were staid from three sets of fish-shaped masts, and rigged square and firm by flat steel rigging. The engine and boiler were put in the car to drive two screw-propellers, right and left-handed, 3 ft. in diameter, with four blades each, occupying three-quarters of the area of the circumference, set at an angle of 60 degrees. A considerable time was spent in perfecting the motive power. Compressed air was tried and abandoned. Tappets, cams, and eccentrics were all tried, to work the slide valve, to obtain the best results. The piston rod of engine passed through both ends of the cylinder, and with long connecting rods worked direct on the crank of the propellers. From memorandum of experiments still preserved the following is a copy of one: June, 27th, 1845, water 50 ozs., spirit 10 ozs., lamp lit 8.45, gauge moves 8.46, engine started 8.48 (100 lb. pressure), engine stopped 8.57, worked 9 minutes, 2,288 revolutions, average 254 per minute. No priming, 40 ozs. water consumed, propulsion (thrust of propellers), 5 lbs. 4 1/2 ozs. at commencement, steady, 4 lbs. 1/2 oz., 57 revolutions to 1 oz. water, steam cut off one-third from beginning.

'The diameter of cylinder of engine was 1 1/2 inch, length of stroke 3 inches.

'In the meantime an engine was also made for the smaller model, and a wing action tried, but with poor results. The time was mostly devoted to the larger model, and in 1847 a tent was erected on Bala Down, about two miles from Chard, and the model taken up one night by the workmen. The experiments were not so favourable as was expected. The machine could not support itself for any distance, but, when launched off, gradually descended, although the power and surface should have been ample; indeed, according to latest calculations, the thrust should have carried more than three times the weight, for there was a thrust of 5 lbs. from the propellers, and a surface of over 70 square feet to sustain under 30 lbs., but necessary speed was lacking.'

Stringfellow himself explained the failure as follows:--

'There stood our aerial protegee in all her purity--too delicate, too fragile, too beautiful for this rough world; at least those were my ideas at the time, but little did I think how soon it was to be realised. I soon found, before I had time to introduce the spark, a drooping in the wings, a flagging in all the parts. In less than ten minutes the machine was saturated with wet from a deposit of dew, so that anything like a trial was impossible by night. I did not consider we could get the silk tight and rigid enough. Indeed, the framework altogether was too weak. The steam-engine was the best part. Our want of success was not for want of power or sustaining surface, but for want of proper adaptation of the means to the end of the various parts.'

Henson, who had spent a considerable amount of money in these experimental constructions, consoled himself for failure by venturing into matrimony; in 1849 he went to America, leaving Stringfellow to continue experimenting alone. From 1846 to 1848 Stringfellow worked on what is really an epoch-making item in the history of aeronautics--the first engine-driven aeroplane which actually flew. The machine in question had a 10 foot span, and was 2 ft. across in the widest part of the wing; the length of tail was 3 ft. 6 ins., and the span of tail in the widest part 22 ins., the total sustaining area being about 14 sq. ft. The motive power consisted of an engine with a cylinder of three-quarter inch diameter and a two-inch stroke; between this and the crank shaft was a bevelled gear giving three revolutions of the propellers to every stroke of the engine; the propellers, right and left screw, were four-bladed and 16 inches in diameter. The total weight of the model with engine was 8 lbs. Its successful flight is ascribed to the fact that Stringfellow curved the wings, giving them rigid front edges and flexible trailing edges, as suggested long before both by Da Vinci and Borelli, but never before put into practice.

Mr F. J. Stringfellow, in the pamphlet quoted above, gives the best account of the flight of this model: 'My father had constructed another small model which was finished early in 1848, and having the loan of a long room in a disused lace factory, early in June the small model was moved there for experiments. The room was about 22 yards long and from 10 to 12 ft. high.... The inclined wire for starting the machine occupied less than half the length of the room and left space at the end for the machine to clear the floor. In the first experiment the tail was set at too high an angle, and the machine rose too rapidly on leaving the wire. After going a few yards it slid back as if coming down an inclined plane, at such an angle that the point of the tail struck the ground and was broken. The tail was repaired and set at a smaller angle. The steam was again got up, and the machine started down the wire, and, upon reaching the point of self-detachment, it gradually rose until it reached the farther end of the room, striking a hole in the canvas placed to stop it. In experiments the machine flew well, when rising as much as one in seven. The late Rev. J. Riste, Esq., lace manufacturer, Northcote Spicer, Esq., J. Toms, Esq., and others witnessed experiments. Mr Marriatt, late of the San Francisco News Letter brought down from London Mr Ellis, the then lessee of Cremorne Gardens, Mr Partridge, and Lieutenant Gale, the aeronaut, to witness experiments. Mr Ellis offered to construct a covered way at Cremorne for experiments. Mr Stringfellow repaired to Cremorne, but not much better accommodations than he had at home were provided, owing to unfulfilled engagement as to room. Mr Stringfellow was preparing for departure when a party of gentlemen unconnected with the Gardens begged to see an experiment, and finding them able to appreciate his endeavours, he got up steam and started the model down the wire. When it arrived at the spot where it should leave the wire it appeared to meet with some obstruction, and threatened to come to the ground, but it soon recovered itself and darted off in as fair a flight as it was possible to make at a distance of about 40 yards, where it was stopped by the canvas.

'Having now demonstrated the practicability of making a steam-engine fly, and finding nothing but a pecuniary loss and little honour, this experimenter rested for a long time, satisfied with what he had effected. The subject, however, had to him special charms, and he still contemplated the renewal of his experiments.'

It appears that Stringfellow's interest did not revive sufficiently for the continuance of the experiments until the founding of the Aeronautical Society of Great Britain in 1866. Wenham's paper on Aerial Locomotion read at the first meeting of the Society, which was held at the Society of Arts under the Presidency of the Duke of Argyll, was the means of bringing Stringfellow back into the field. It was Wenham's suggestion, in the first place, that monoplane design should be abandoned for the superposition of planes; acting on this suggestion Stringfellow constructed a model triplane, and also designed a steam engine of slightly over one horse-power, and a one horse-power copper boiler and fire box which, although capable of sustaining a pressure of 500 lbs. to the square inch, weighed only about 40 lbs.

Both the engine and the triplane model were exhibited at the first Aeronautical Exhibition held at the Crystal Palace in 1868. The triplane had a supporting surface of 28 sq. ft.; inclusive of engine, boiler, fuel, and water its total weight was under 12 lbs. The engine worked two 21 in. propellers at 600 revolutions per minute, and developed 100 lbs. steam pressure in five minutes, yielding one-third horse-power. Since no free flight was allowed in the Exhibition, owing to danger from fire, the triplane was suspended from a wire in the nave of the building, and it was noted that, when running along the wire, the model made a perceptible lift.

A prize of L100 was awarded to the steam engine as the lightest steam engine in proportion to its power. The engine and model together may be reckoned as Stringfellow's best achievement. He used his L100 in preparation for further experiments, but he was now an old man, and his work was practically done. Both the triplane and the engine were eventually bought for the Washington Museum; Stringfellow's earlier models, together with those constructed by him in conjunction with Henson, remain in this country in the Victoria and Albert Museum.

John Stringfellow died on December 13th, 1883. His place in the history of aeronautics is at least equal to that of Cayley, and it may be said that he laid the foundation of such work as was subsequently accomplished by Maxim, Langley, and their fellows. It was the coming of the internal combustion engine that rendered flight practicable, and had this prime mover been available in John Stringfellow's day the Wright brothers' achievement might have been antedated by half a century.

V. WENHAM, LE BRIS, AND SOME OTHERS

There are few outstanding events in the development of aeronautics between Stringfellow's final achievement and the work of such men as Lilienthal, Pilcher, Montgomery, and their kind; in spite of this, the later middle decades of the nineteenth century witnessed a considerable amount of spade work both in England and in France, the two countries which led in the way in aeronautical development until Lilienthal gave honour to Germany, and Langley and Montgomery paved the way for the Wright Brothers in America.

Two abortive attempts characterised the sixties of last century in France. As regards the first of these, it was carried out by three men, Nadar, Ponton d'Amecourt, and De la Landelle, who conceived the idea of a full-sized helicopter machine. D'Amecourt exhibited a steam model, constructed in 1865, at the Aeronautical Society's Exhibition in 1868. The engine was aluminium with cylinders of bronze, driving two screws placed one above the other and rotating in Opposite directions, but the power was not sufficient to lift the model. De la Landelle's principal achievement consisted in the publication in 1863 of a book entitled Aviation which has a certain historical value; he got out several designs for large machines on the helicopter principle, but did little more until the three combined in the attempt to raise funds for the construction of their full-sized machine. Since the funds were not forthcoming, Nadar took to ballooning as the means of raising money; apparently he found this substitute for real flight sufficiently interesting to divert him from the study of the helicopter principle, for the experiment went no further.

The other experimenter of this period, one Count d'Esterno, took out a patent in 1864 for a soaring machine which allowed for alteration of the angle of incidence of the wings in the manner that was subsequently carried out by the Wright Brothers. It was not until 1883 that any attempt was made to put this patent to practical use, and, as the inventor died while it was under construction, it was never completed. D'Esterno was also responsible for the production of a work entitled Du Vol des Oiseaux, which is a very remarkable study of the flight of birds.

Mention has already been made of the founding of the Aeronautical Society of Great Britain, which, since 1918 has been the Royal Aeronautical Society. 1866 witnessed the first meeting of the Society under the Presidency of the Duke of Argyll, when in June, at the Society of Arts, Francis Herbert Wenham read his now classic paper Aerial Locomotion. Certain quotations from this will show how clearly Wenham had thought out the problems connected with flight.

'The first subject for consideration is the proportion of surface to weight, and their combined effect in descending perpendicularly through the atmosphere. The datum is here based upon the consideration of safety, for it may sometimes be needful for a living being to drop passively, without muscular effort. One square foot of sustaining surface for every pound of the total weight will be sufficient for security.

'According to Smeaton's table of atmospheric resistances, to produce a force of one pound on a square foot, the wind must move against the plane (or which is the same thing, the plane against the wind), at the rate of twenty-two feet per second, or 1,320 feet per minute, equal to fifteen miles per hour. The resistance of the air will now balance the weight on the descending surface, and, consequently, it cannot exceed that speed. Now, twenty-two feet per second is the velocity acquired at the end of a fall of eight feet--a height from which a well-knit man or animal may leap down without much risk of injury. Therefore, if a man with parachute weigh together 143 lbs., spreading the same number of square feet of surface contained in a circle fourteen and a half feet in diameter, he will descend at perhaps an unpleasant velocity, but with safety to life and limb.

'It is a remarkable fact how this proportion of wing-surface to weight extends throughout a great variety of the flying portion of the animal kingdom, even down to hornets, bees, and other insects. In some instances, however, as in the gallinaceous tribe, including pheasants, this area is somewhat exceeded, but they are known to be very poor fliers. Residing as they do chiefly on the ground, their wings are only required for short distances, or for raising them or easing their descent from their roosting-places in forest trees, the shortness of their wings preventing them from taking extended flights. The wing-surface of the common swallow is rather more than in the ratio of two square feet per pound, but having also great length of pinion, it is both swift and enduring in its flight. When on a rapid course this bird is in the habit of furling its wings into a narrow compass. The greater extent of surface is probably needful for the continual variations of speed and instant stoppages for obtaining its insect food.

'On the other hand, there are some birds, particularly of the duck tribe, whose wing-surface but little exceeds half a square foot, or seventy-two inches per pound, yet they may be classed among the strongest and swiftest of fliers. A weight of one pound, suspended from an area of this extent, would acquire a velocity due to a fall of sixteen feet--a height sufficient for the destruction or injury of most animals. But when the plane is urged forward horizontally, in a manner analogous to the wings of a bird during flight, the sustaining power is greatly influenced by the form and arrangement of the surface.

'In the case of perpendicular descent, as a parachute, the sustaining effect will be much the same, whatever the figure of the outline of the superficies may be, and a circle perhaps affords the best resistance of any. Take, for example, a circle of twenty square feet (as possessed by the pelican) loaded with as many pounds. This, as just stated, will limit the rate of perpendicular descent to 1,320 feet per minute. But instead of a circle sixty-one inches in diameter, if the area is bounded by a parallelogram ten feet long by two feet broad, and whilst at perfect freedom to descend perpendicularly, let a force be applied exactly in a horizontal direction, so as to carry it edgeways, with the long side foremost, at a forward speed of thirty miles per hour--just double that of its passive descent: the rate of fall under these conditions will be decreased most remarkably, probably to less than one-fifteenth part, or eighty-eight feet per minute, or one mile per hour.'

And again: 'It has before been shown how utterly inadequate the mere perpendicular impulse of a plane is found to be in supporting a weight, when there is no horizontal motion at the time. There is no material weight of air to be acted upon, and it yields to the slightest force, however great the velocity of impulse may be. On the other hand, suppose that a large bird, in full flight, can make forty miles per hour, or 3,520 feet per minute, and performs one stroke per second. Now, during every fractional portion of that stroke, the wing is acting upon and obtaining an impulse from a fresh and undisturbed body of air; and if the vibration of the wing is limited to an arc of two feet, this by no means represents the small force of action that would be obtained when in a stationary position, for the impulse is secured upon a stratum of fifty-eight feet in length of air at each stroke. So that the conditions of weight of air for obtaining support equally well apply to weight of air and its reaction in producing forward impulse.

'So necessary is the acquirement of this horizontal speed, even in commencing flight, that most heavy birds, when possible, rise against the wind, and even run at the top of their speed to make their wings available, as in the example of the eagle, mentioned at the commencement of this paper. It is stated that the Arabs, on horseback, can approach near enough to spear these birds, when on the plain, before they are able to rise; their habit is to perch on an eminence, where possible.

'The tail of a bird is not necessary for flight. A pigeon can fly perfectly with this appendage cut short off; it probably performs an important function in steering, for it is to be remarked, that most birds that have either to pursue or evade pursuit are amply provided with this organ.

'The foregoing reasoning is based upon facts, which tend to show that the flight of the largest and heaviest of all birds is really performed with but a small amount of force, and that man is endowed with sufficient muscular power to enable him also to take individual and extended flights, and that success is probably only involved in a question of suitable mechanical adaptations. But if the wings are to be modelled in imitation of natural examples, but very little consideration will serve to demonstrate its utter impracticability when applied in these forms.'

Thus Wenham, one of the best theorists of his age. The Society with which this paper connects his name has done work, between that time and the present, of which the importance cannot be overestimated, and has been of the greatest value in the development of aeronautics, both in theory and experiment. The objects of the Society are to give a stronger impulse to the scientific study of aerial navigation, to promote the intercourse of those interested in the subject at home and abroad, and to give advice and instruction to those who study the principles upon which aeronautical science is based. From the date of its foundation the Society has given special study to dynamic flight, putting this before ballooning. Its library, its bureau of advice and information, and its meetings, all assist in forwarding the study of aeronautics, and its twenty-three early Annual Reports are of considerable value, containing as they do a large amount of useful information on aeronautical subjects, and forming practically the basis of aeronautical science.

Ante to Wenham, Stringfellow and the French experimenters already noted, by some years, was Le Bris, a French sea captain, who appears to have required only a thorough scientific training to have rendered him of equal moment in the history of gliding flight with Lilienthal himself. Le Bris, it appears, watched the albatross and deduced, from the manner in which it supported itself in the air, that plane surfaces could be constructed and arranged to support a man in like manner. Octave Chanute, himself a leading exponent of gliding, gives the best description of Le Bris's experiments in a work, Progress in Flying Machines, which, although published as recently as I 1894, is already rare. Chanute draws from a still rarer book, namely, De la Landelle's work published in 1884. Le Bris himself, quoted by De la Landelle as speaking of his first visioning of human flight, describes how he killed an albatross, and then--'I took the wing of the albatross and exposed it to the breeze; and lo! in spite of me it drew forward into the wind; notwithstanding my resistance it tended to rise. Thus I had discovered the secret of the bird! I comprehended the whole mystery of flight.'

This apparently took place while at sea; later on Le Bris, returning to France, designed and constructed an artificial albatross of sufficient size to bear his own weight. The fact that he followed the bird outline as closely as he did attests his lack of scientific training for his task, while at the same time the success of the experiment was proof of his genius. The body of his artificial bird, boat-shaped, was 13

1/2 ft. in length, with a breadth of 4 ft. at the widest part. The material was cloth stretched over a wooden framework; in front was a small mast rigged after the manner of a ship's masts to which were attached poles and cords with which Le Bris intended to work the wings. Each wing was 23 ft. in length, giving a total supporting surface of nearly 220 sq. ft.; the weight of the whole apparatus was only 92 pounds. For steering, both vertical and horizontal, a hinged tail was provided, and the leading edge of each wing was made flexible. In construction throughout, and especially in that of the wings, Le Bris adhered as closely as possible to the original albatross.

He designed an ingenious kind of mechanism which he termed 'Rotules,' which by means of two levers gave a rotary motion to the front edge of the wings, and also permitted of their adjustment to various angles. The inventor's idea was to stand upright in the body of the contrivance, working the levers and cords with his hands, and with his feet on a pedal by means of which the steering tail was to be worked. He anticipated that, given a strong wind, he could rise into the air after the manner of an albatross, without any need for flapping his wings, and the account of his first experiment forms one of the most interesting incidents in the history of flight. It is related in full in Chanute's work, from which the present account is summarised.

Le Bris made his first experiment on a main road near Douarnenez, at Trefeuntec. From his observation of the albatross Le Bris concluded that it was necessary to get some initial velocity in order to make the machine rise; consequently on a Sunday morning, with a breeze of about 12 miles an hour blowing down the road, he had his albatross placed on a cart and set off, with a peasant driver, against the wind. At the outset the machine was fastened to the cart by a rope running through the rails on which the machine rested, and secured by a slip knot on Le Bris's own wrist, so that only a jerk on his part was necessary to loosen the rope and set the machine free. On each side walked an assistant holding the wings, and when a turn of the road brought the machine full into the wind these men were instructed to let go, while the driver increased the pace from a walk to a trot. Le Bris, by pressure on the levers of the machine, raised the front edges of his wings slightly; they took the wind almost instantly to such an extent that the horse, relieved of a great part of the weight he had been drawing, turned his trot into a gallop. Le Bris gave the jerk of the rope that should have unfastened the slip knot, but a concealed nail on the cart caught the rope, so that it failed to run. The lift of the machine was such, however, that it relieved the horse of very nearly the weight of the cart and driver, as well as that of Le Bris and his machine, and in the end the rails of the cart gave way. Le Bris rose in the air, the machine maintaining perfect balance and rising to a height of nearly 300 ft., the total length of the glide being upwards of an eighth of a mile. But at the last moment the rope which had originally fastened the machine to the cart got wound round the driver's body, so that this unfortunate dangled in the air under Le Bris and probably assisted in maintaining the balance of the artificial albatross. Le Bris, congratulating himself on his success, was prepared to enjoy just as long a time in the air as the pressure of the wind would permit, but the howls of the unfortunate driver at the end of the rope beneath him dispelled his dreams; by working his levers he altered the angle of the front wing edges so skilfully as to make a very successful landing indeed for the driver, who, entirely uninjured, disentangled himself from the rope as soon as he touched the ground, and ran off to retrieve his horse and cart.

Apparently his release made a difference in the centre of gravity, for Le Bris could not manipulate his levers for further ascent; by skilful manipulation he retarded the descent sufficiently to escape injury to himself; the machine descended at an angle, so that one wing, striking the ground in front of the other, received a certain amount of damage.

It may have been on account of the reluctance of this same or another driver that Le Bris chose a different method of launching himself in making a second experiment with his albatross. He chose the edge of a quarry which had been excavated in a depression of the ground; here he assembled his apparatus at the bottom of the quarry, and by means of a rope was hoisted to a height of nearly 100 ft. from the quarry bottom, this rope being attached to a mast which he had erected upon the edge of the depression in which the quarry was situated. Thus hoisted, the albatross was swung to face a strong breeze that blew inland, and Le Bris manipulated his levers to give the front edges of his wings a downward angle, so that only the top surfaces should take the wing pressure. Having got his balance, he obtained a lifting angle of incidence on the wings by means of his levers, and released the hook that secured the machine, gliding off over the quarry. On the glide he met with the inevitable upward current of air that the quarry and the depression in which it was situated caused; this current upset the balance of the machine and flung it to the bottom of the quarry, breaking it to fragments. Le Bris, apparently as intrepid as ingenious, gripped the mast from which his levers were worked, and, springing upward as the machine touched earth, escaped with no more damage than a broken leg. But for the rebound of the levers he would have escaped even this.

The interest of these experiments is enhanced by the fact that Le Bris was a seafaring man who conducted them from love of the science which had fired his imagination, and in so doing exhausted his own small means. It was in 1855 that he made these initial attempts, and twelve years passed before his persistence was rewarded by a public subscription made at Brest for the purpose of enabling him to continue his experiments. He built a second albatross, and on the advice of his friends ballasted it for flight instead of travelling in it himself. It was not so successful as the first, probably owing to the lack of human control while in flight; on one of the trials a height of 150 ft. was attained, the glider being secured by a thin rope and held so as to face into the wind. A glide of nearly an eighth of a mile was made with the rope hanging slack, and, at the end of this distance, a rise in the ground modified the force of the wind, whereupon the machine settled down without damage. A further trial in a gusty wind resulted in the complete destruction of this second machine; Le Bris had no more funds, no further subscriptions were likely to materialise, and so the experiments of this first exponent of the art of gliding (save for Besnier and his kind) came to an end. They constituted a notable achievement, and undoubtedly Le Bris deserves a better place than has been accorded him in the ranks of the early experimenters.

Contemporary with him was Charles Spencer, the first man to practice gliding in England. His apparatus consisted of a pair of wings with a total area of 30 sq. ft., to which a tail and body were attached. The weight of this apparatus was some 24 lbs., and, launching himself on it from a small eminence, as was done later by Lilienthal in his experiments, the inventor made flights of over 120 feet. The glider in question was exhibited at the Aeronautical Exhibition of 1868.

VI. THE AGE OF THE GIANTS

Until the Wright Brothers definitely solved the problem of flight and virtually gave the aeroplane its present place in aeronautics, there were three definite schools of experiment. The first of these was that which sought to imitate nature by means of the ornithopter or flapping-wing machines directly imitative of bird flight; the second school was that which believed in the helicopter or lifting screw; the third and eventually successful school is that which followed up the principle enunciated by Cayley, that of opposing a plane surface to the resistance of the air by supplying suitable motive power to drive it at the requisite angle for support.

Engineering problems generally go to prove that too close an imitation of nature in her forms of recipro-cating motion is not advantageous; it is impossible to copy the minutiae of a bird's wing effectively, and the bird in flight depends on the tiniest details of its feathers just as much as on the general principle on which the whole wing is constructed. Bird flight, however, has attracted many experimenters, including even Lilienthal; among others may be mentioned F. W. Brearey, who invented what he called the 'Pectoral cord,' which stored energy on each upstroke of the artificial wing; E. P. Frost; Major R. Moore, and especially Hureau de Villeneuve, a most enthusiastic student of this form of flight, who began his experiments about 1865, and altogether designed and made nearly 300 artificial birds, one of his later constructions was a machine in bird form with a wing span of about 50 ft.; the motive power for this was supplied by steam from a boiler which, being stationary on the ground, was connected by a length of hose to the machine. De Villeneuve, turning on steam for his first trial, obtained sufficient power to make the wings beat very forcibly; with the inventor on the machine the latter rose several feet into the air, whereupon de Villeneuve grew nervous and turned off the steam supply. The machine fell to the earth, breaking one of its wings, and it does not appear that de Villeneuve troubled to reconstruct it. This experiment remains as the greatest success yet achieved by any machine constructed on the ornithopter principle.

It may be that, as forecasted by the prophet Wells, the flapping-wing machine will yet come to its own and compete with the aeroplane in efficiency. Against this, however, are the practical advantages of the rotary mechanism of the aeroplane propeller as compared with the movement of a bird's wing, which, according to Marey, moves in a figure of eight. The force derived from a propeller is of necessity continual, while it is equally obvious that that derived from a flapping movement is intermittent, and, in the recovery of a wing after completion of one stroke for the next, there is necessarily a certain cessation, if not loss, of power.

The matter of experiment along any lines in connection with aviation is primarily one of hard cash. Throughout the whole history of flight up to the outbreak of the European war development has been handicapped on the score of finance, and, since the arrival of the aeroplane, both ornithopter and helicopter schools have been handicapped by this consideration. Thus serious study of the efficiency of wings in imitation of those of the living bird has not been carried to a point that might win success for this method of propulsion. Even Wilbur Wright studied this subject and propounded certain theories, while a later and possibly more scientific student, F. W. Lanchester, has also contributed empirical conclusions. Another and earlier student was Lawrence Hargrave, who made a wing-propelled model which achieved successful flight, and in 1885 was exhibited before the Royal Society of New South Wales. Hargrave called the principle on which his propeller worked that of a 'Trochoided plane'; it was, in effect, similar to the feathering of an oar.

Hargrave, to diverge for a brief while from the machine to the man, was one who, although he achieved nothing worthy of special remark, contributed a great deal of painstaking work to the science of flight. He made a series of experiments with man-lifting kites in addition to making a study of flapping-wing flight. It cannot be said that he set forth any new principle; his work was mainly imitative, but at the same time by developing ideas originated in great measure by others he helped toward the solution of the problem.

Attempts at flight on the helicopter principle consist in the work of De la Landelle and others already mentioned. The possibility of flight by this method is modified by a very definite disadvantage of which lovers of the helicopter seem to take little account. It is always claimed for a machine of this type that it possesses great advantages both in rising and in landing, since, if it were effective, it would obviously be able to rise from and alight on any ground capable of containing its own bulk; a further advantage claimed is that the helicopter would be able to remain stationary in the air, maintaining itself in any position by the vertical lift of its propeller.

These potential assets do not take into consideration the fact that efficiency is required not only in rising, landing, and remaining stationary in the air, but also in actual flight. It must be evident that if a certain amount of the motive force is used in maintaining the machine off the ground, that amount of force is missing from the total of horizontal driving power. Again, it is often assumed by advocates of this form of flight that the rapidity of climb of the helicopter would be far greater than that of the driven plane; this view overlooks the fact that the maintenance of aerodynamic support would claim the greater part of the engine-power; the rate of ascent would be governed by the amount of power that could be developed surplus to that required for maintenance.

This is best explained by actual figures: assuming that a propeller 15 ft. in diameter is used, almost 50 horse-power would be required to get an upward lift of 1,000 pounds; this amount of horse-power would be continually absorbed in maintaining the machine in the air at any given level; for actual lift from one level to another at a speed of eleven feet per second a further 20 horse-power would be required, which means that 70 horse-power must be constantly provided for; this absorption of power in the mere maintenance of aero-dynamic support is a permanent drawback.

The attraction of the helicopter lies, probably, in the ease with which flight is demonstrated by means of models constructed on this principle, but one truism with regard to the principles of flight is that the problems change remarkably, and often unexpectedly, with the size of the machine constructed for experiment. Berriman, in a brief but very interesting manual entitled Principles of Flight, assumed that 'there is a significant dimension of which the effective area is an expression of the second power, while the weight became an expression of the third power. Then once again we have the two-thirds power law militating against the successful construction of large helicopters, on the ground that the essential weight increases disproportionately fast to the effective area. From a consideration of the structural features of propellers it is evident that this particular relationship does not apply in practice, but it seems reasonable that some such governing factor should exist as an explanation of the apparent failure of all full-sized machines that have been constructed. Among models there is nothing more strikingly successful than the toy helicopter, in which the essential weight is so small compared with the effective area.'

De la Landelle's work, already mentioned, was carried on a few years later by another Frenchman, Castel, who constructed a machine with eight propellers arranged in two fours and driven by a compressed air motor or engine. The model with which Castel experimented had a total weight of only 49 lbs.; it rose in the air and smashed itself by driving against a wall, and the inventor does not seem to have proceeded further. Contemporary with Castel was Professor Forlanini, whose design was for a machine very similar to de la Landelle's, with two superposed screws. This machine ranks as the second on the helicopter principle to achieve flight; it remained in the air for no less than the third of a minute in one of its trials.

Later experimenters in this direction were Kress, a German; Professor Wellner, an Austrian; and W. R. Kimball, an American. Kress, like most Germans, set to the development of an idea which others had originated; he followed de la Landelle and Forlanini by fitting two superposed propellers revolving in opposite directions, and with this machine he achieved good results as regards horse-power to weight; Kimball, it appears, did not get beyond the rubber-driven model stage, and any success he may have achieved was modified by the theory enunciated by Berriman and quoted above.

Comparing these two schools of thought, the helicopter and bird-flight schools, it appears that the latter has the greater chance of eventual success--that is, if either should ever come into competition with the aeroplane as effective means of flight. So far, the aeroplane holds the field, but the whole science of flight is so new and so full of unexpected developments that this is no reason for assuming that other means may not give equal effect, when money and brains are diverted from the driven plane to a closer imitation of natural flight.

Reverting from non-success to success, from consideration of the two methods mentioned above to the direction in which practical flight has been achieved, it is to be noted that between the time of Le Bris, Stringfellow, and their contemporaries, and the nineties of last century, there was much plodding work carried out with little visible result, more especially so far as English students were concerned. Among the incidents of those years is one of the most pathetic tragedies in the whole history of aviation, that of Alphonse Penaud, who, in his thirty years of life, condensed the experience of his predecessors and combined it with his own genius to state in a published patent what the aeroplane of to-day should be. Consider the following abstract of Penaud's design as published in his patent of 1876, and comparison of this with the aeroplane that now exists will show very few divergences except for those forced on the inventor by the fact that the internal combustion engine had not then developed. The double surfaced planes were to be built with wooden ribs and arranged with a slight dihedral angle; there was to be a large aspect ratio and the wings were cambered as in Stringfellow's later models. Provision was made for warping the wings while in flight, and the trailing edges were so designed as to be capable of upward twist while the machine was in the air. The planes were to be placed above the car, and provision was even made for a glass wind-screen to give protection to the pilot during flight. Steering was to be accomplished by means of lateral and vertical planes forming a tail; these controlled by a single lever corresponding to the 'joy stick' of the present day plane.

Penaud conceived this machine as driven by two propellers; alternatively these could be driven by petrol or steam-fed motor, and the centre of gravity of the machine while in flight was in the front fifth of the wings. Penaud estimated from 20 to 30 horse-power sufficient to drive this machine, weighing with pilot and passenger 2,600 lbs., through the air at a speed of 60 miles an hour, with the wings set at an angle of incidence of two degrees. So complete was the design that it even included instruments, consisting of an aneroid, pressure indicator, an anemometer, a compass, and a level. There, with few alterations, is the aeroplane as we know it--and Penaud was twenty-seven when his patent was published.

For three years longer he worked, experimenting with models, contributing essays and other valuable data to French papers on the subject of aeronautics. His gains were ill health, poverty, and neglect, and at the age of thirty a pistol shot put an end to what had promised to be one of the most brilliant careers in all the history of flight.

Two years before the publication of Penaud's patent Thomas Moy experimented at the Crystal Palace with a twin-propelled aeroplane, steam driven, which seems to have failed mainly because the internal combustion engine had not yet come to give sufficient power for weight. Moy anchored his machine to a pole running on a prepared circular track; his engine weighed 80 lbs. and, developing only three horse-power, gave him a speed of 12 miles an hour. He himself estimated that the machine would not rise until he could get a speed of 35 miles an hour, and his estimate was correct. Two six-bladed propellers were placed side by side between the two main planes of the machine, which was supported on a triangular wheeled undercarriage and steered by fairly conventional tail planes. Moy realised that he could not get sufficient power to achieve flight, but he went on experimenting in various directions, and left much data concerning his experiments which has not yet been deemed worthy of publication, but which still contains a mass of information that is of practical utility, embodying as it does a vast amount of painstaking work.

Penaud and Moy were followed by Goupil, a Frenchman, who, in place of attempting to fit a motor to an aeroplane, experimented by making the wind his motor. He anchored his machine to the ground, allowing it two feet of lift, and merely waited for a wind to come along and lift it. The machine was stream lined, and the wings, curving as in the early German patterns of war aeroplanes, gave a total lifting surface of about 290 sq. ft. Anchored to the ground and facing a wind of 19 feet per second, Goupil's machine lifted its own weight and that of two men as well to the limit of its anchorage. Although this took place as late as 1883 the inventor went no further in practical work. He published a book, however, entitled La Locomotion Aerienne, which is still of great importance, more especially on the subject of inherent stability.

In 1884 came the first patents of Horatio Phillips, whose work lay mainly in the direction of investigation into the curvature of plane surfaces, with a view to obtaining the greatest amount of support. Phillips was one of the first to treat the problem of curvature of planes as a matter for scientific experiment, and, great as has been the development of the driven plane in the 36 years that have passed since he began, there is still room for investigation into the subject which he studied so persistently and with such valuable result.

At this point it may be noted that, with the solitary exception of Le Bris, practically every student of flight had so far set about constructing the means of launching humanity into the air without any attempt at ascertaining the nature and peculiarities of the sustaining medium. The attitude of experimenters in general might be compared to that of a man who from boyhood had grown up away from open water, and, at the first sight of an expanse of water, set to work to construct a boat with a vague idea that, since wood would float, only sufficient power was required to make him an efficient navigator. Accident, perhaps, in the shape of lack of means of procuring driving power, drove Le Bris to the form of experiment which he actually carried out; it remained for the later years of the nineteenth century to produce men who were content to ascertain the nature of the support the air would afford before attempting to drive themselves through it.

Of the age in which these men lived and worked, giving their all in many cases to the science they loved, even to life itself, it may be said with truth that 'there were giants on the earth in those days,' as far as aeronautics is in question. It was an age of giants who lived and dared and died, venturing into uncharted space, knowing nothing of its dangers, giving, as a man gives to his mistress, without stint and for the joy of the giving. The science of to-day, compared with the glimmerings that were in that age of the giants, is a fixed and certain thing; the problems of to-day are minor problems, for the great major problem vanished in solution when the Wright Brothers made their first ascent. In that age of the giants was evolved the flying man, the new type in human species which found full expression and came to full development in the days of the war, achieving feats of daring and endurance which leave the commonplace landsman staggered at thought of that of which his fellows prove themselves capable. He is a new type, this flying man, a being of self-forgetfulness; of such was Lilienthal, of such was Pilcher; of such in later days were Farman, Bleriot, Hamel, Rolls, and their fellows; great names that will live for as long as man flies, adventurers equally with those of the spacious days of Elizabeth. To each of these came the call, and he worked and dared and passed, having, perhaps, advanced one little step in the long march that has led toward the perfecting of flight.

It is not yet twenty years since man first flew, but into that twenty years have been compressed a century or so of progress, while, in the two decades that preceded it, was compressed still more. We have only to recall and recount the work of four men: Lilienthal, Langley, Pilcher, and Clement Ader to see the immense stride that was made between the time when Penaud pulled a trigger for the last time and the Wright Brothers first left the earth. Into those two decades was compressed the investigation that meant knowledge of the qualities of the air, together with the development of the one prime mover that rendered flight a possibility--the internal combustion engine. The coming and progress of this latter is a thing apart, to be detailed separately; for the present we are concerned with the evolution of the driven plane, and with it the evolution of that daring being, the flying man. The two are inseparable, for the men gave themselves to their art; the story of Lilienthal's life and death is the story of his work; the story of Pilcher's work is that of his life and death.

Considering the flying man as he appeared in the war period, there entered into his composition a new element--patriotism--which brought about a modification of the type, or, perhaps, made it appear that certain men belonged to the type who in reality were commonplace mortals, animated, under normal conditions, by normal motives, but driven by the stress of the time to take rank with the last expression of human energy, the flying type. However that may be, what may be termed the mathematising of aeronautics has rendered the type itself evanescent; your pilot of to-day knows his craft, once he is trained, much in the manner that a driver of a motor-lorry knows his vehicle; design has been systematised, capabilities have been tabulated; camber, dihedral angle, aspect ratio, engine power, and plane surface, are business items of drawing office and machine shop; there is room for enterprise, for genius, and for skill; once and again there is room for daring, as in the first Atlantic flight. Yet that again was a thing of mathematical calculation and petrol storage, allied to a certain stark courage which may be found even in landsmen. For the ventures into the unknown, the limit of daring, the work for work's sake, with the almost certainty that the final reward was death, we must look back to the age of the giants, the age when flying was not a business, but romance.

VII. LILIENTHAL AND PILCHER

There was never a more enthusiastic and consistent student of the problems of flight than Otto Lilienthal, who was born in 1848 at Anklam, Pomerania, and even from his early school-days dreamed and planned the conquest of the air. His practical experiments began when, at the age of thirteen, he and his brother Gustav made wings consisting of wooden framework covered with linen, which Otto attached to his arms, and then ran downhill flapping them. In consequence of possible derision on the part of other boys, Otto confined these experiments for the most part to moonlit nights, and gained from them some idea of the resistance offered by flat surfaces to the air. It was in 1867 that the two brothers began really practical work, experimenting with wings which, from their design, indicate some knowledge of Besnier and the history of his gliding experiments; these wings the brothers fastened to their backs, moving them with their legs after the fashion of one attempting to swim. Before they had achieved any real success in gliding the Franco-German war came as an interruption; both brothers served in this campaign, resuming their experiments in 1871 at the conclusion of hostilities.

The experiments made by the brothers previous to the war had convinced Otto that previous experimenters in gliding flight had failed through reliance on empirical conclusions or else through incomplete observation on their own part, mostly of bird flight. From 1871 onward Otto Lilenthal (Gustav's interest in the problem was not maintained as was his brother's) made what is probably the most detailed and accurate series of observations that has ever been made with regard to the properties of curved wing surfaces. So far as could be done, Lilienthal tabulated the amount of air resistance offered to a bird's wing, ascertaining that the curve is necessary to flight, as offering far more resistance than a flat surface. Cayley, and others, had already stated this, but to Lilienthal belongs the honour of being first to put the statement to effective proof--he made over 2,000 gliding flights between 1891 and the regrettable end of his experiments; his practical conclusions are still regarded as part of the accepted theory of students of flight. In 1889 he published a work on the subject of gliding flight which stands as data for investigators, and, on the conclusions embodied in this work, he began to build his gliders and practice what he had preached, turning from experiment with models to wings that he could use.

It was in the summer of 1891 that he built his first glider of rods of peeled willow, over which was stretched strong cotton fabric; with this, which had a supporting surface of about 100 square feet, Otto Lilienthal launched himself in the air from a spring board, making glides which, at first of only a few feet, gradually lengthened. As his experience of the supporting qualities of the air progressed he gradually altered his designs until, when Pilcher visited him in the spring of 1895, he experimented with a glider, roughly made of peeled willow rods and cotton fabric, having an area of 150 square feet and weighing half a hundredweight. By this time Lilienthal had moved from his springboard to a conical artificial hill which he had had thrown up on level ground at Grosse Lichterfelde, near Berlin. This hill was made with earth taken from the excavations incurred in constructing a canal, and had a cave inside in which Lilienthal stored his machines. Pilcher, in his paper on 'Gliding,'[1] gives an excellent short summary of Lilienthal's experiments, from which the following extracts are taken:--

'In 1892 a canal was being cut, close to where Lilienthal lived, in the suburbs of Berlin, and with the surplus earth Lilienthal had a special hill thrown up to fly from. The country round is as flat as the sea, and there is not a house or tree near it to make the wind unsteady, so this was an ideal practicing ground; for practicing on natural hills is generally rendered very difficult by shifty and gusty winds.... This hill is 50 feet high, and conical. Inside the hill there is a cave for the machines to be kept in.... When Lilienthal made a good flight he used to land 300 feet from the centre of the hill, having come down at an angle of 1 in 6; but his best flights have been at an angle of about 1 in 10.

'If it is calm, one must run a few steps down the hill, holding the machine as far back on oneself as possible, when the air will gradually support one, and one slides off the hill into the air. If there is any wind, one should face it at starting; to try to start with a side wind is most unpleasant. It is possible after a great deal of practice to turn in the air, and fairly quickly. This is accomplished by throwing one's weight to one side, and thus lowering the machine on that side towards which one wants to turn. Birds do the same thing--crows and gulls show it very clearly. Last year Lilienthal chiefly experimented with double-surfaced machines. These were very much like the old machines with awnings spread above them.

'The object of making these double-surfaced machines was to get more surface without increasing the length and width of the machine. This, of course, it does, but I personally object to any machine in which the wing surface is high above

---

1 Aeronautical Classes, No. 5. Royal Aeronautical Society's publications. 'At first Lilienthal used to experiment by jumping off a springboard with a good run. Then he took to practicing on some hills close to Berlin. In the summer of 1892 he built a flat-roofed hut on the summit of a hill, from the top of which he used to jump, trying, of course, to soar as far as possible before landing.... One of the great dangers with a soaring machine is losing forward speed, inclining the machine too much down in front, and coming down head first. Lilienthal was the first to introduce the system of handling a machine in the air merely by moving his weight about in the machine; he always rested only on his elbows or on his elbows and shoulders....

the weight. I consider that it makes the machine very difficult to handle in bad weather, as a puff of wind striking the surface, high above one, has a great tendency to heel the machine over.

'Herr Lilienthal kindly allowed me to sail down his hill in one of these double-surfaced machines last June. With the great facility afforded by his conical hill the machine was handy enough; but I am afraid I should not be able to manage one at all in the squally districts I have had to practice in over here.

'Herr Lilienthal came to grief through deserting his old method of balancing. In order to control his tipping movements more rapidly he attached a line from his horizontal rudder to his head, so that when he moved his head forward it would lift the rudder and tip the machine up in front, and vice versa. He was practicing this on some natural hills outside Berlin, and he apparently got muddled with the two motions, and, in trying to regain speed after he had, through a lull in the wind, come to rest in the air, let the machine get too far down in front, came down head first and was killed.'

Then in another passage Pilcher enunciates what is the true value of such experiments as Lilienthal--and, subsequently, he himself--made: 'The object of experimenting with soaring machines,' he says, 'is to enable one to have practice in starting and alighting and controlling a machine in the air. They cannot possibly float horizontally in the air for any length of time, but to keep going must necessarily lose in elevation. They are excellent schooling machines, and that is all they are meant to be, until power, in the shape of an engine working a screw propeller, or an engine working wings to drive the machine forward, is added; then a person who is used to soaring down a hill with a simple soaring machine will be able to fly with comparative safety. One can best compare them to bicycles having no cranks, but on which one could learn to balance by coming down an incline.'

It was in 1895 that Lilienthal passed from experiment with the monoplane type of glider to the construction of a biplane glider which, according to his own account, gave better results than his previous machines. 'Six or seven metres velocity of wind,' he says, 'sufficed to enable the sailing surface of 18 square metres to carry me almost horizontally against the wind from the top of my hill without any starting jump. If the wind is stronger I allow myself to be simply lifted from the point of the hill and to sail slowly towards the wind. The direction of the flight has, with strong wind, a strong upwards tendency. I often reach positions in the air which are much higher than my starting point. At the climax of such a line of flight I sometimes come to a standstill for some time, so that I am enabled while floating to speak with the gentlemen who wish to photograph me, regarding the best position for the photographing.'

Lilienthal's work did not end with simple gliding, though he did not live to achieve machine-driven flight. Having, as he considered, gained sufficient experience with gliders, he constructed a power-driven machine which weighed altogether about 90 lbs., and this was thoroughly tested. The extremities of its wings were made to flap, and the driving power was obtained from a cylinder of compressed carbonic acid gas, released through a hand-operated valve which, Lilienthal anticipated, would keep the machine in the air for four minutes. There were certain minor accidents to the mechanism, which delayed the trial flights, and on the day that Lilienthal had determined to make his trial he made a long gliding flight with a view to testing a new form of rudder that--as Pilcher relates--was worked by movements of his head. His death came about through the causes that Pilcher states; he fell from a height of 50 feet, breaking his spine, and the next day he died.

It may be said that Lilienthal accomplished as much as any one of the great pioneers of flying. As brilliant in his conceptions as da Vinci had been in his, and as conscientious a worker as Borelli, he laid the foundations on which Pilcher, Chanute, and Professor Montgomery were able to build to such good purpose. His book on bird flight, published in 1889, with the authorship credited both to Otto and his brother Gustav, is regarded as epoch-making; his gliding experiments are no less entitled to this description.

In England Lilienthal's work was carried on by Percy Sinclair Pilcher, who, born in 1866, completed six years' service in the British Navy by the time that he was nineteen, and then went through a course of engineering, subsequently joining Maxim in his experimental work. It was not until 1895 that he began to build the first of the series of gliders with which he earned his plane among the pioneers of flight. Probably the best account of Pilcher's work is that given in the Aeronautical Classics issued by the Royal Aeronautical Society, from which the following account of Pilcher's work is mainly abstracted[2].

2 Aeronautical Classes, No. 5. Royal Aeronautical Society publications.

The 'Bat,' as Pilcher named his first glider, was a monoplane which he completed before he paid his visit to Lilienthal in 1895. Concerning this Pilcher stated that he purposely finished his own machine before going to see Lilienthal, so as to get the greatest advantage from any original ideas he might have; he was not able to make any trials with this machine, however, until after witnessing Lilienthal's experiments and making several glides in the biplane glider which Lilienthal constructed.

The wings of the 'Bat' formed a pronounced dihedral angle; the tips being raised 4 feet above the body. The spars forming the entering edges of the wings crossed each other in the centre and were lashed to opposite sides of the triangle that served as a mast for the stay-wires that guyed the wings. The four ribs of each wing, enclosed in pockets in the fabric, radiated fanwise from the centre, and were each stayed by three steel piano-wires to the top of the triangular mast, and similarly to its base. These ribs were bolted down to the triangle at their roots, and could be easily folded back on to the body when the glider was not in use. A small fixed vertical surface was carried in the rear. The framework and ribs were made entirely of Riga pine; the surface fabric was nainsook. The area of the machine was 150 square feet; its weight 45 lbs.; so that in flight, with Pilcher's weight of 145 lbs. added, it carried one and a half pounds to the square foot.

Pilcher's first glides, which he carried out on a grass hill on the banks of the Clyde near Cardross, gave little result, owing to the exaggerated dihedral angle of the wings, and the absence of a horizontal tail. The 'Bat 'was consequently reconstructed with a horizontal tail plane added to the vertical one, and with the wings lowered so that the tips were only six inches above the level of the body. The machine now gave far better results; on the first glide into a head wind Pilcher rose to a height of twelve feet and remained in the the air for a third of a minute; in the second attempt a rope was used to tow the glider, which rose to twenty feet and did not come to earth again until nearly a minute had passed. With experience Pilcher was able to lengthen his glide and improve his balance, but the dropped wing tips made landing difficult, and there were many breakages.

In consequence of this Pilcher built a second glider which he named the 'Beetle,' because, as he said, it looked like one. In this the square-cut wings formed almost a continuous plane, rigidly fixed to the central body, which consisted of a shaped girder. These wings were built up of five transverse bamboo spars, with two shaped ribs running from fore to aft of each wing, and were stayed overhead to a couple of masts. The tail, consisting of two discs placed crosswise (the horizontal one alone being movable), was carried high up in the rear. With the exception of the wing-spars, the whole framework was built of white pine. The wings in this machine were actually on a higher level than the operator's head; the centre of gravity was, consequently, very low, a fact which, according to Pilcher's own account, made the glider very difficult to handle. Moreover, the weight of the 'Beetle,' 80 lbs., was considerable; the body had been very solidly built to enable it to carry the engine which Pilcher was then contemplating; so that the glider carried some 225 lbs. with its area of 170 square feet--too great a mass for a single man to handle with comfort.

It was in the spring of 1896 that Pilcher built his third glider, the 'Gull,' with 300 square feet of area and a weight of 55 lbs. The size of this machine rendered it unsuitable for experiment in any but very calm weather, and it incurred such damage when experiments were made in a breeze that Pilcher found it necessary to build a fourth, which he named the 'Hawk.' This machine was very soundly built, being constructed of bamboo, with the exception of the two main transverse beams. The wings were attached to two vertical masts, 7 feet high, and 8 feet apart, joined at their summits and their centres by two wooden beams. Each wing had nine bamboo ribs, radiating from its mast, which was situated at a distance of 2 feet 6 inches from the forward edge of the wing. Each rib was rigidly stayed at the top of the mast by three tie-wires, and by a similar number to the bottom of the mast, by which means the curve of each wing was maintained uniformly. The tail was formed of a triangular horizontal surface to which was affixed a triangular vertical surface, and was carried from the body on a high bamboo mast, which was also stayed from the masts by means of steel wires, but only on its upper surface, and it was the snapping of one of these guy wires which caused the collapse of the tail support and brought about the fatal end of Pilcher's experiments. In flight, Pilcher's head, shoulders, and the greater part of his chest projected above the wings. He took up his position by passing his head and shoulders through the top aperture formed between the two wings, and resting his forearms on the longitudinal body members. A very simple form of undercarriage, which took the weight off the glider on the ground, was fitted, consisting of two bamboo rods with wheels suspended on steel springs.

Balance and steering were effected, apart from the high degree of inherent stability afforded by the tail, as in the case of Lilienthal's glider, by altering the position of the body. With this machine Pilcher made some twelve glides at Eynsford in Kent in the summer of 1896, and as he progressed he increased the length of his glides, and also handled the machine more easily, both in the air and in landing. He was occupied with plans for fitting an engine and propeller to the 'Hawk,' but, in these early days of the internal combustion engine, was unable to get one light enough for his purpose. There were rumours of an engine weighing 15 lbs. which gave 1 horse-power, and was reported to be in existence in America, but it could not be traced.

In the spring of 1897 Pilcher took up his gliding experiments again, obtaining what was probably the best of his glides on June 19th, when he alighted after a perfectly balanced glide of over 250 yards in length, having crossed a valley at a considerable height. From his various experiments he concluded that once the machine was launched in the air an engine of, at most, 3 horse-power would suffice for the maintenance of horizontal flight, but he had to allow for the additional weight of the engine and propeller, and taking into account the comparative inefficiency of the propeller, he planned for an engine of 4 horse-power. Engine and propeller together were estimated at under 44 lbs. weight, the engine was to be fitted in front of the operator, and by means of an overhead shaft was to operate the propeller situated in rear of the wings. 1898 went by while this engine was under construction. Then in 1899 Pilcher became interested in Lawrence Hargrave's soaring kites, with which he carried out experiments during the summer of 1899. It is believed that he intended to incorporate a number of these kites in a new machine, a triplane, of which the fragments remaining are hardly sufficient to reconstitute the complete glider. This new machine was never given a trial. For on September 30th, 1899, at Stamford Hall, Market Harborough, Pilcher agreed to give a demonstration of gliding flight, but owing to the unfavourable weather he decided to postpone the trial of the new machine and to experiment with the 'Hawk,' which was intended to rise from a level field, towed by a line passing over a tackle drawn by two horses. At the first trial the machine rose easily, but the tow-line snapped when it was well clear of the ground, and the glider descended, weighed down through being sodden with rain. Pilcher resolved on a second trial, in which the glider again rose easily to about thirty feet, when one of the guy wires of the tail broke, and the tail collapsed; the machine fell to the ground, turning over, and Pilcher was unconscious when he was freed from the wreckage.

Hopes were entertained of his recovery, but he died on Monday, October 2nd, 1899, aged only thirty-four. His work in the cause of flying lasted only four years, but in that time his actual accomplishments were sufficient to place his name beside that of Lilienthal, with whom he ranks as one of the greatest exponents of gliding flight.

VIII. AMERICAN GLIDING EXPERIMENTS

While Pilcher was carrying on Lilienthal's work in England, the great German had also a follower in America; one Octave Chanute, who, in one of the statements which he has left on the subject of his experiments acknowledges forty years' interest in the problem of flight, did more to develop the glider in America than--with the possible exception of Montgomery--any other man. Chanute had all the practicality of an American; he began his work, so far as actual gliding was concerned, with a full-sized glider of the Lilienthal type, just before Lilienthal was killed. In a rather rare monograph, entitled Experiments in Flying, Chanute states that he found the Lilienthal glider hazardous and decided to test the value of an idea of his own; in this he followed the same general method, but reversed the principle upon which Lilienthal had depended for maintaining his equilibrium in the air. Lilienthal had shifted the weight of his body, under immovable wings, as fast and as far as the sustaining pressure varied under his surfaces; this shifting was mainly done by moving the feet, as the actions required were small except when alighting. Chanute's idea was to have the operator remain seated in the machine in the air, and to intervene only to steer or to alight; moving mechanism was provided to adjust the wings automatically in order to restore balance when necessary.

Chanute realised that experiments with models were of little use; in order to be fully instructive, these experiments should be made with a full-sized machine which carried its operator, for models seldom fly twice alike in the open air, and no relation can be gained from them of the divergent air currents which they have experienced. Chanute's idea was that any flying machine which might be constructed must be able to operate in a wind; hence the necessity for an operator to report upon what occurred in flight, and to acquire practical experience of the work of the human factor in imitation of bird flight. From this point of view he conducted his own experiments; it must be noted that he was over sixty years of age when he began, and, being no longer sufficiently young and active to perform any but short and insignificant glides, the courage of the man becomes all the more noteworthy; he set to work to evolve the state required by the problem of stability, and without any expectation of advancing to the construction of a flying machine which might be of commercial value. His main idea was the testing of devices to secure equilibrium; for this purpose he employed assistants to carry out the practical work, where he himself was unable to supply the necessary physical energy.

Together with his assistants he found a suitable place for experiments among the sandhills on the shore of Lake Michigan, about thirty miles eastward from Chicago. Here a hill about ninety-five feet high was selected as a point from which Chanute's gliders could set off; in practice, it was found that the best observation was to be obtained from short glides at low speed, and, consequently, a hill which was only sixty-one feet above the shore of the lake was employed for the experimental work done by the party.

In the years 1896 and 1897, with parties of from four to six persons, five full-sized gliders were tried out, and from these two distinct types were evolved: of these one was a machine consisting of five tiers of wings and a steering tail, and the other was of the biplane type; Chanute believed these to be safer than any other machine previously evolved, solving, as he states in his monograph, the problem of inherent equilibrium as fully as this could be done. Unfortunately, very few photographs were taken of the work in the first year, but one view of a multiple wing-glider survives, showing the machine in flight. In 1897 a series of photographs was taken exhibiting the consecutive phases of a single flight; this series of photographs represents the experience gained in a total of about one thousand glides, but the point of view was varied so as to exhibit the consecutive phases of one single flight.

The experience gained is best told in Chanute's own words. 'The first thing,' he says, 'which we discovered practically was that the wind flowing up a hill-side is not a steadily-flowing current like that of a river. It comes as a rolling mass, full of tumultuous whirls and eddies, like those issuing from a chimney; and they strike the apparatus with constantly varying force and direction, sometimes withdrawing support when most needed. It has long been known, through instrumental observations, that the wind is constantly changing in force and direction; but it needed the experience of an operator afloat on a gliding machine to realise that this all proceeded from cyclonic action; so that more was learned in this respect in a week than had previously been acquired by several years of experiments with models. There was a pair of eagles, living in the top of a dead tree about two miles from our tent, that came almost daily to show us how such wind effects are overcome and utilised. The birds swept in circles overhead on pulseless wings, and rose high up in the air. Occasionally there was a side-rocking motion, as of a ship rolling at sea, and then the birds rocked back to an even keel; but although we thought the action was clearly automatic, and were willing to learn, our teachers were too far off to show us just how it was done, and we had to experiment for ourselves.'

Chanute provided his multiple glider with a seat, but, since each glide only occupied between eight and twelve seconds, there was little possibility of the operator seating himself. With the multiple glider a pair of horizontal bars provided rest for the arms, and beyond these was a pair of vertical bars which the operator grasped with his hands; beyond this, the operator was in no way attached to the machine. He took, at the most, four running steps into the wind, which launched him in the air, and thereupon he sailed into the wind on a generally descending course. In the matter of descent Chanute observed the sparrow and decided to imitate it. 'When the latter,' he says, 'approaches the street, he throws his body back, tilts his outspread wings nearly square to the course, and on the cushion of air thus encountered he stops his speed and drops lightly to the ground. So do all birds. We tried it with misgivings, but found it perfectly effective. The soft sand was a great advantage, and even when the experts were racing there was not a single sprained ankle.'

With the multiple winged glider some two to three hundred glides were made without any accident either to the man or to the machine, and the action was found so effective, the principle so sound, that full plans were published for the benefit of any experimenters who might wish to improve on this apparatus. The American

Aeronautical Annual for 1897 contains these plans; Chanute confessed that some movement on the part of the operator was still required to control the machine, but it was only a seventh or a sixth part of the movement required for control of the Lilienthal type.

Chanute waxed enthusiastic over the possibilities of gliding, concerning which he remarks that 'There is no more delightful sensation than that of gliding through the air. All the faculties are on the alert, and the motion is astonishingly smooth and elastic. The machine responds instantly to the slightest movement of the operator; the air rushes by one's ears; the trees and bushes flit away underneath, and the landing comes all too quickly. Skating, sliding, and bicycling are not to be compared for a moment to aerial conveyance, in which, perhaps, zest is added by the spice of danger. For it must be distinctly understood that there is constant danger in such preliminary experiments. When this hazard has been eliminated by further evolution, gliding will become a most popular sport.'

Later experiments proved that the biplane type of glider gave better results than the rather cumbrous model consisting of five tiers of planes. Longer and more numerous glides, to the number of seven to eight hundred, were obtained, the rate of descent being about one in six. The longest distance traversed was about 120 yards, but Chanute had dreams of starting from a hill about 200 feet high, which would have given him gliding flights of 1,200 feet. He remarked that 'In consequence of the speed gained by running, the initial stage of the flight is nearly horizontal, and it is thrilling to see the operator pass from thirty to forty feet overhead, steering his machine, undulating his course, and struggling with the wind-gusts which whistle through the guy wires. The automatic mechanism restores the angle of advance when compromised by variations of the breeze; but when these come from one side and tilt the apparatus, the weight has to be shifted to right the machine... these gusts sometimes raise the machine from ten to twenty feet vertically, and sometimes they strike the apparatus from above, causing it to descend suddenly. When sailing near the ground, these vicissitudes can be counteracted by movements of the body from three to four inches; but this has to be done instantly, for neither wings nor gravity will wait on meditation. At a height of three hundred or four hundred feet the regulating mechanism would probably take care of these wind-gusts, as it does, in fact, for their minor variations. The speed of the machine is generally about seventeen miles an hour over the ground, and from twenty-two to thirty miles an hour relative to the air. Constant effort was directed to keep down the velocity, which was at times fifty-two miles an hour. This is the purpose of the starting and gliding against the wind, which thus furnishes an initial velocity without there being undue speed at the landing. The highest wind we dared to experiment in blew at thirty-one miles an hour; when the wind was stronger, we waited and watched the birds.'

Chanute details an amusing little incident which occurred in the course of experiment with the biplane glider. He says that 'We had taken one of the machines to the top of the hill, and loaded its lower wings with sand to hold it while we e went to lunch. A gull came strolling inland, and flapped full-winged to inspect. He swept several circles above the machine, stretched his neck, gave a squawk and went off. Presently he returned with eleven other gulls, and they seemed to hold a conclave about one hundred feet above the big new white bird which they had discovered on the sand. They circled round after round, and once in a while there was a series of loud peeps, like those of a rusty gate, as if in conference, with sudden flutterings, as if a terrifying suggestion had been made. The bolder birds occasionally swooped downwards to inspect the monster more closely; they twisted their heads around to bring first one eye and then the other to bear, and then they rose again. After some seven or eight minutes of this performance, they evidently concluded either that the stranger was too formidable to tackle, if alive, or that he was not good to eat, if dead, and they flew off to resume fishing, for the weak point about a bird is his stomach.'

The gliders were found so stable, more especially the biplane form, that in the end Chanute permitted amateurs to make trials under guidance, and throughout the whole series of experiments not a single accident occurred. Chanute came to the conclusion that any young, quick, and handy man could master a gliding machine almost as soon as he could get the hang of a bicycle, although the penalty for any mistake would be much more severe.

At the conclusion of his experiments he decided that neither the multiple plane nor the biplane type of glider was sufficiently perfected for the application of motive power. In spite of the amount of automatic stability that he had obtained he considered that there was yet more to be done, and he therefore advised that every possible method of securing stability and safety should be tested, first with models, and then with full-sized machines; designers, he said, should make a point of practice in order to make sure of the action, to proportion and adjust the parts of their machine, and to eliminate hidden defects. Experimental flight, he suggested, should be tried over water, in order to break any accidental fall; when a series of experiments had proved the stability of a glider, it would then be time to apply motive power. He admitted that such a process would be both costly and slow, but, he said, that 'it greatly diminished the chance of those accidents which bring a whole line of investigation into contempt.' He saw the flying machine as what it has, in fact, been; a child of evolution, carried on step by step by one investigator after another, through the stages of doubt and perplexity which lie behind the realm of possibility, beyond which is the present day stage of actual performance and promise of ultimate success and triumph over the earlier, more cumbrous, and slower forms of the transport that we know.

Chanute's monograph, from which the foregoing notes have been comprised, was written soon after the conclusion of his series of experiments. He does not appear to have gone in for further practical work, but to have studied the subject from a theoretical view-point and with great attention to the work done by others. In a paper contributed in 1900 to the American Independent, he remarks that 'Flying machines promise better results as to speed, but yet will be of limited commercial application. They may carry mails and reach other inaccessible places, but they cannot compete with railroads as carriers of passengers or freight. They will not fill the heavens with commerce, abolish custom houses, or revolutionise the world, for they will be expensive for the loads which they can carry, and subject to too many weather contingencies. Success is, however, probable. Each experimenter has added something to previous knowledge which his successors can avail of. It now seems likely that two forms of flying machines, a sporting type and an exploration type, will be gradually evolved within one or two generations, but the evolution will be costly and slow, and must be carried on by well-equipped and thoroughly informed scientific men; for the casual inventor, who relies upon one or two happy inspirations, will have no chance of success whatever.'

Follows Professor John J. Montgomery, who, in the true American spirit, describes his own experiments so well that nobody can possibly do it better. His account of his work was given first of all in the American Journal, Aeronautics, in

January, 1909, and thence transcribed in the English paper of the same name in May, 1910, and that account is here copied word for word. It may, however, be noted first that as far back as 1860, when Montgomery was only a boy, he was attracted to the study of aeronautical problems, and in 1883 he built his first machine, which was of the flapping-wing ornithopter type, and which showed its designer, with only one experiment, that he must design some other form of machine if he wished to attain to a successful flight. Chanute details how, in 1884 and 1885 Montgomery built three gliders, demonstrating the value of curved surfaces. With the first of these gliders Montgomery copied the wing of a seagull; with the second he proved that a flat surface was virtually useless, and with the third he pivoted his wings as in the Antoinette type of power-propelled aeroplane, proving to his own satisfaction that success lay in this direction. His own account of the gliding flights carried out under his direction is here set forth, being the best description of his work that can be obtained:--

'When I commenced practical demonstration in my work with aeroplanes I had before me three points; first, equilibrium; second, complete control; and third, long continued or soaring flight. In starting I constructed and tested three sets of models, each in advance of the other in regard to the continuance of their soaring powers, but all equally perfect as to equilibrium and control. These models were tested by dropping them from a cable stretched between two mountain tops, with various loads, adjustments and positions. And it made no difference whether the models were dropped upside down or any other conceivable position, they always found their equilibrium immediately and glided safely to earth.

'Then I constructed a large machine patterned after the first model, and with the assistance of three cowboy friends personally made a number of flights in the steep mountains near San Juan (a hundred miles distant). In making these flights I simply took the aeroplane and made a running jump. These tests were discontinued after I put my foot into a squirrel hole in landing and hurt my leg.

'The following year I commenced the work on a larger scale, by engaging aeronauts to ride my aeroplane dropped from balloons. During this work I used five hot-air balloons and one gas balloon, five or six aeroplanes, three riders--Maloney, Wilkie, and Defolco--and had sixteen applicants on my list, and had a training station to prepare any when I needed them.

'Exhibitions were given in Santa Cruz, San Jose, Santa Clara, Oaklands, and Sacramento. The flights that were made, instead of being haphazard affairs, were in the order of safety and development. In the first flight of an aeronaut the aeroplane was so arranged that the rider had little liberty of action, consequently he could make only a limited flight. In some of the first flights, the aeroplane did little more than settle in the air. But as the rider gained experience in each successive flight I changed the adjustments, giving him more liberty of action, so he could obtain longer flights and more varied movements in the flights. But in none of the flights did I have the adjustments so that the riders had full liberty, as I did not consider that they had the requisite knowledge and experience necessary for their safety; and hence, none of my aeroplanes were launched so arranged that the rider could make adjustments necessary for a full flight.

'This line of action caused a good deal of trouble with aeronauts or riders, who had unbounded confidence and wanted to make long flights after the first few trials; but I found it necessary, as they seemed slow in comprehending the important elements and were willing to take risks. To give them the full knowledge in these matters I was formulating plans for a large starting station on the Mount Hamilton Range from which I could launch an aeroplane capable of carrying two, one of my aeronauts and myself, so I could teach him by demonstration. But the disasters consequent on the great earthquake completely stopped all my work on these lines. The flights that were given were only the first of the series with aeroplanes patterned after the first model. There were no aeroplanes constructed according to the two other models, as I had not given the full demonstration of the workings of the first, though some remarkable and startling work was done. On one occasion Maloney, in trying to make a very short turn in rapid flight, pressed very hard on the stirrup which gives a screw-shape to the wings, and made a side somersault. The course of the machine was very much like one turn of a corkscrew. After this movement the machine continued on its regular course. And afterwards Wilkie, not to be outdone by Maloney, told his friends he would do the same, and in a subsequent flight made two side somersaults, one in one direction and the other in an opposite, then made a deep dive and a long glide, and, when about three hundred feet in the air, brought the aeroplane to a sudden stop and settled to the earth. After these antics, I decreased the extent of the possible change in the form of wing-surface, so as to allow only straight sailing or only long curves in turning.

'During my work I had a few carping critics that I silenced by this standing offer: If they would deposit a thousand dollars I would cover it on this proposition. I would fasten a 150 pound sack of sand in the rider's seat, make the necessary adjustments, and send up an aeroplane upside down with a balloon, the aeroplane to be liberated by a time fuse. If the aeroplane did not immediately right itself, make a flight, and come safely to the ground, the money was theirs.

'Now a word in regard to the fatal accident. The circumstances are these: The ascension was given to entertain a military company in which were many of Maloney's friends, and he had told them he would give the most sensational flight they ever heard of. As the balloon was rising with the aeroplane, a guy rope dropping switched around the right wing and broke the tower that braced the two rear wings and which also gave control over the tail. We shouted Maloney that the machine was broken, but he probably did not hear us, as he was at the same time saying, "Hurrah for Montgomery's airship," and as the break was behind him, he may not have detected it. Now did he know of the breakage or not, and if he knew of it did he take a risk so as not to disappoint his friends? At all events, when the machine started on its flight the rear wings commenced to flap (thus indicating they were loose), the machine turned on its back, and settled a little faster than a parachute. When we reached Maloney he was unconscious and lived only thirty minutes. The only mark of any kind on him was a scratch from a wire on the side of his neck. The six attending physicians were puzzled at the cause of his death. This is remarkable for a vertical descent of over 2,000 feet.'

The flights were brought to an end by the San Francisco earthquake in April, 1906, which, Montgomery states, 'Wrought such a disaster that I had to turn my attention to other subjects and let the aeroplane rest for a time.' Montgomery resumed experiments in 1911 in California, and in October of that year an accident brought his work to an end. The report in the American Aeronautics says that 'a little whirlwind caught the machine and dashed it head on to the ground; Professor Montgomery landed on his head and right hip. He did not believe himself seriously hurt, and talked with his year-old bride in the tent. He complained of pains in his back, and continued to grow worse until he died.'

IX. NOT PROVEN

The early history of flying, like that of most sciences, is replete with tragedies; in addition to these it contains one mystery concerning Clement Ader, who was well known among European pioneers in the development of the telephone, and first turned his attention to the problems of mechanical flight in 1872. At the outset he favoured the ornithopter principle, constructing a machine in the form of a bird with a wing-spread of twenty-six feet; this, according to Ader's conception, was to fly through the efforts of the operator. The result of such an attempt was past question and naturally the machine never left the ground.

A pause of nineteen years ensued, and then in 1886 Ader turned his mind to the development of the aeroplane, constructing a machine of bat-like form with a wingspread of about forty-six feet, a weight of eleven hundred pounds, and a steam-power plant of between twenty and thirty horse-power driving a four-bladed tractor screw. On October 9th, 1890, the first trials of this machine were made, and it was alleged to have flown a distance of one hundred and sixty-four feet. Whatever truth there may be in the allegation, the machine was wrecked through deficient equilibrium at the end of the trial. Ader repeated the construction, and on October 14th, 1897, tried out his third machine at the military establishment at Satory in the presence of the French military authorities, on a circular track specially prepared for the experiment. Ader and his friends alleged that a flight of nearly a thousand feet was made; again the machine was wrecked at the end of the trial, and there Ader's practical work may be said to have ended, since no more funds were forthcoming for the subsidy of experiments.

There is the bald narrative, but it is worthy of some amplification. If Ader actually did what he claimed, then the position which the Wright Brothers hold as first to navigate the air in a power-driven plane is nullified. Although at this time of writing it is not a quarter of a century since Ader's experiment in the presence of witnesses competent to judge on his accomplishment, there is no proof either way, and whether he was or was not the first man to fly remains a mystery in the story of the conquest of the air.

The full story of Ader's work reveals a persistence and determination to solve the problem that faced him which was equal to that of Lilienthal. He began by penetrating into the interior of Algeria after having disguised himself as an Arab, and there he spent some months in studying flight as practiced by the vultures of the district. Returning to France in 1886 he began to construct the 'Eole,' modelling it, not on the vulture, but in the shape of a bat. Like the Lilienthal and Pilcher gliders this machine was fitted with wings which could be folded; the first flight made, as already noted, on October 9th, 1890, took place in the grounds of the chateau d'Amainvilliers, near Bretz; two fellow-enthusiasts named Espinosa and Vallier stated that a flight was actually made; no statement in the history of aeronautics has been subject of so much question, and the claim remains unproved.

It was in September of 1891 that Ader, by permission of the Minister of War, moved the 'Eole' to the military establishment at Satory for the purpose of further trial. By this time, whether he had flown or not, his nineteen years of work in connection with the problems attendant on mechanical flight had attracted so much attention that henceforth his work was subject to the approval of the military authorities, for already it was recognised that an efficient flying machine would confer an inestimable advantage on the power that possessed it in the event of war. At Satory the 'Eole' was alleged to have made a flight of 109 yards, or, according to another account, 164 feet, as stated above, in the trial in which the machine wrecked itself through colliding with some carts which had been placed near the track--the root cause of this accident, however, was given as deficient equilibrium.

Whatever the sceptics may say, there is reason for belief in the accomplishment of actual flight by Ader with his first machine in the fact that, after the inevitable official delay of some months, the French War Ministry granted funds for further experiment. Ader named his second machine, which he began to build in May, 1892, the 'Avion,' and--an honour which he well deserve--that name remains in French aeronautics as descriptive of the power-driven aeroplane up to this day.

This second machine, however, was not a success, and it was not until 1897 that the second 'Avion,' which was the third power-driven aeroplane of Ader's construction, was ready for trial. This was fitted with two steam motors of twenty horse-power each, driving two four-bladed propellers; the wings warped automatically: that is to say, if it were necessary to raise the trailing edge of one wing on the turn, the trailing edge of the opposite wing was also lowered by the same movement; an under-carriage was also fitted, the machine running on three small wheels, and levers controlled by the feet of the aviator actuated the movement of the tail planes.

On October the 12th, 1897, the first trials of this 'Avion' were made in the presence of General Mensier, who admitted that the machine made several hops above the ground, but did not consider the performance as one of actual flight. The result was so encouraging, in spite of the partial failure, that, two days later, General Mensier, accompanied by General Grillon, a certain Lieutenant Binet, and two civilians named respectively Sarrau and Leaute, attended for the purpose of giving the machine an official trial, over which the great controversy regarding Ader's success or otherwise may be said to have arisen.

We will take first Ader's own statement as set out in a very competent account of his work published in Paris in 1910. Here are Ader's own words: 'After some turns of the propellers, and after travelling a few metres, we started off at a lively pace; the pressure-gauge registered about seven atmospheres; almost immediately the vibrations of the rear wheel ceased; a little later we only experienced those of the front wheels at intervals. 'Unhappily, the wind became suddenly strong, and we had some difficulty in keeping the "Avion" on the white line. We increased the pressure to between eight and nine atmospheres, and immediately the speed increased considerably, and the vibrations of the wheels were no longer sensible; we were at that moment at the point marked G in the sketch; the "Avion" then found itself freely supported by its wings; under the impulse of the wind it continually tended to go outside the (prepared) area to the right, in spite of the action of the rudder. On reaching the point V it found itself in a very critical position; the wind blew strongly and across the direction of the white line which it ought to follow; the machine then, although still going forward, drifted quickly out of the area; we immediately put over the rudder to the left as far as it would go; at the same time increasing the pressure still more, in order to try to regain the course. The "Avion" obeyed, recovered a little, and remained for some seconds headed towards its intended course, but it could not struggle against the wind; instead of going back, on the contrary it drifted farther and farther away. And ill-luck had it that the drift took the direction towards part of the School of Musketry, which was guarded by posts and barriers. Frightened at the prospect of breaking ourselves against these obstacles, surprised at seeing the earth getting farther away from under the "Avion," and very much impressed by seeing it rushing sideways at a sickening speed, instinctively we stopped everything. What passed through our thoughts at this moment which threatened a tragic turn would be difficult to set down. All at once came a great shock, splintering, a heavy concussion: we had landed.'

Thus speaks the inventor; the cold official mind gives out a different account, crediting the 'Avion' with merely a few hops, and to-day, among those who consider the problem at all, there is a little group which persists in asserting that to Ader belongs the credit of the first power-driven flight, while a larger group is equally persistent in stating that, save for a few ineffectual hops, all three wheels of the machine never left the ground. It is past question that the 'Avion' was capable of power-driven flight; whether it achieved it or no remains an unsettled problem.

Ader's work is negative proof of the value of such experiments as Lilienthal, Pilcher, Chanute, and Montgomery conducted; these four set to work to master the eccentricities of the air before attempting to use it as a supporting medium for continuous flight under power; Ader attacked the problem from the other end; like many other experimenters he regarded the air as a stable fluid capable of giving such support to his machine as still water might give to a fish, and he reckoned that he had only to produce the machine in order to achieve flight. The wrecked 'Avion' and the refusal of the French War Ministry to grant any more funds for further experiment are sufficient evidence of the need for working along the lines taken by the pioneers of gliding rather than on those which Ader himself adopted.

Let it not be thought that in this comment there is any desire to derogate from the position which Ader should occupy in any study of the pioneers of aeronautical enterprise. If he failed, he failed magnificently, and if he succeeded, then the student of aeronautics does him an injustice and confers on the Brothers Wright an honour which, in spite of the value of their work, they do not deserve. There was one earlier than Ader, Alphonse Penaud, who, in the face of a lesser disappointment than that which Ader must have felt in gazing on the wreckage of his machine, committed suicide; Ader himself, rendered unable to do more, remained content with his achievement, and with the knowledge that he had played a good part in the long search which must eventually end in triumph. Whatever the world might say, he himself was certain that he had achieved flight. This, for him, was perforce enough.

Before turning to consideration of the work accomplished by the Brothers Wright, and their proved conquest of the air, it is necessary first to sketch as briefly as may be the experimental work of Sir (then Mr) Hiram Maxim, who, in his book, Artificial and Natural Flight, has given a fairly complete account of his various experiments. He began by experimenting with models, with screw-propelled planes so attached to a horizontal movable arm that when the screw was set in motion the plane described a circle round a central point, and, eventually, he built a giant aeroplane having a total supporting area of 1,500 square feet, and a wing-span of fifty feet. It has been thought advisable to give a fairly full description of the power plant used to the propulsion of this machine in the section devoted to engine development. The aeroplane, as Maxim describes it, had five long and narrow planes projecting from each side, and a main or central plane of pterygoid aspect. A fore and aft rudder was provided, and had all the auxiliary planes been put in position for experimental work a total lifting surface of 6,000 square feet could have been obtained. Maxim, however, did not use more than 4,000 square feet of lifting surface even in his later experiments; with this he judged the machine capable of lifting slightly under 8,000 lbs. weight, made up of 600 lbs. water in the boiler and tank, a crew of three men, a supply of naphtha fuel, and the weight of the machine itself.

Maxim's intention was, before attempting free flight, to get as much data as possible regarding the conditions under which flight must be obtained, by what is known in these days as 'taxi-ing'--that is, running the propellers at sufficient speed to drive the machine along the ground without actually mounting into the air. He knew that he had an immense lifting surface and a tremendous amount of power in his engine even when the total weight of the experimental plant was taken into consideration, and thus he set about to devise some means of keeping the machine on the nine foot gauge rail track which had been constructed for the trials. At the outset he had a set of very heavy cast-iron wheels made on which to mount the machine, the total weight of wheels, axles, and connections being about one and a half tons. These were so constructed that the light flanged wheels which supported the machine on the steel rails could be lifted six inches above the track, still leaving the heavy wheels on the rails for guidance of the machine. 'This arrangement,'

Maxim states, 'was tried on several occasions, the machine being run fast enough to lift the forward end off the track. However, I found considerable difficulty in starting and stopping quickly on account of the great weight, and the amount of energy necessary to set such heavy wheels spinning at a high velocity. The last experiment with these wheels was made when a head wind was blowing at the rate of about ten miles an hour. It was rather unsteady, and when the machine was running at its greatest velocity, a sudden gust lifted not only the front end, but also the heavy front wheels completely off the track, and the machine falling on soft ground was soon blown over by the wind.'

Consequently, a safety track was provided, consisting of squared pine logs, three inches by nine inches, placed about two feet above the steel way and having a thirty-foot gauge. Four extra wheels were fitted to the machine on outriggers and so adjusted that, if the machine should lift one inch clear of the steel rails, the wheels at the ends of the outriggers would engage the under side of the pine trackway.

The first fully loaded run was made in a dead calm with 150 lbs. steam pressure to the square inch, and there was no sign of the wheels leaving the steel track. On a second run, with 230 lbs. steam pressure the machine seemed to alternate between adherence to the lower and upper tracks, as many as three of the outrigger wheels engaging at the same time, and the weight on the steel rails being reduced practically to nothing. In preparation for a third run, in which it was intended to use full power, a dynamometer was attached to the machine and the engines were started at 200 lbs. pressure, which was gradually increased to 310 lbs per square inch. The incline of the track, added to the reading of the dynamometer, showed a total screw thrust of 2,164 lbs. After the dynamometer test had been completed, and everything had been made ready for trial in motion, careful observers were stationed on each side of the track, and the order was given to release the machine. What follows is best told in Maxim's own words:--

'The enormous screw-thrust started the engine so quickly that it nearly threw the engineers off their feet, and the machine bounded over the track at a great rate. Upon noticing a slight diminution in the steam pressure, I turned on more gas, when almost instantly the steam commenced to blow a steady blast from the small safety valve, showing that the pressure was at least 320 lbs. in the pipes supplying the engines with steam. Before starting on this run, the wheels that were to engage the upper track were painted, and it was the duty of one of my assistants to observe these wheels during the run, while another assistant watched the pressure gauges and dynagraphs. The first part of the track was up a slight incline, but the machine was lifted clear of the lower rails and all of the top wheels were fully engaged on the upper track when about 600 feet had been covered. The speed rapidly increased, and when 900 feet had been covered, one of the rear axle trees, which were of two-inch steel tubing, doubled up and set the rear end of the machine completely free. The pencils ran completely across the cylinders of the dynagraphs and caught on the underneath end. The rear end of the machine being set free, raised considerably above the track and swayed. At about 1,000 feet, the left forward wheel also got clear of the upper track, and shortly afterwards the right forward wheel tore up about 100 feet of the upper track. Steam was at once shut off and the machine sank directly to the earth, embedding the wheels in the soft turf without leaving any other marks, showing most conclusively that the machine was completely suspended in the air before it settled to the earth. In this accident, one of the pine timbers forming the upper track went completely through the lower framework of the machine and broke a number of the tubes, but no damage was done to the machinery except a slight injury to one of the screws.'

It is a pity that the multifarious directions in which Maxim turned his energies did not include further development of the aeroplane, for it seems fairly certain that he was as near solution of the problem as Ader himself, and, but for the holding-down outer track, which was really the cause of his accident, his machine would certainly have achieved free flight, though whether it would have risen, flown and alighted, without accident, is matter for conjecture.

The difference between experiments with models and with full-sized machines is emphasised by Maxim's statement to the effect that with a small apparatus for ascertaining the power required for artificial flight, an angle of incidence of one in fourteen was most advantageous, while with a large machine he found it best to increase his angle to one in eight in order to get the maximum lifting effect on a short run at a moderate speed. He computed the total lifting effect in the experiments which led to the accident as not less than 10,000 lbs., in which is proof that only his rail system prevented free flight.

X. SAMUEL PIERPOINT LANGLEY

Langley was an old man when he began the study of aeronautics, or, as he himself might have expressed it, the study of aerodromics, since he persisted in calling the series of machines he built 'Aerodromes,' a word now used only to denote areas devoted to use as landing spaces for flying machines; the Wright Brothers, on the other hand, had the great gift of youth to aid them in their work. Even so it was a great race between Langley, aided by Charles Manly, and Wilbur and Orville Wright, and only the persistent ill-luck which dogged Langley from the start to the finish of his experiments gave victory to his rivals. It has been proved conclusively in these later years of accomplished flight that the machine which Langley launched on the Potomac River in October of 1903 was fully capable of sustained flight, and only the accidents incurred in launching prevented its pilot from being the first man to navigate the air successfully in a power-driven machine.

The best account of Langley's work is that diffused throughout a weighty tome issued by the Smithsonian Institution, entitled the Langley Memoir on Mechanical Flight, of which about one-third was written by Langley himself, the remainder being compiled by Charles M. Manly, the engineer responsible for the construction of the first radial aero-engine, and chief assistant to Langley in his experiments. To give a twentieth of the contents of this volume in the present short account of the development of mechanical flight would far exceed the amount of space that can be devoted even to so eminent a man in aeronautics as S. P. Langley, who, apart from his achievement in the construction of a power-driven aeroplane really capable of flight, was a scientist of no mean order, and who brought to the study of aeronautics the skill of the trained investigator allied to the inventive resource of the genius.

That genius exemplified the antique saw regarding the infinite capacity for taking pains, for the Langley Memoir shows that as early as 1891 Langley had completed a set of experiments, lasting through years, which proved it possible to construct machines giving such a velocity to inclined surfaces that bodies indefinitely heavier than air could be sustained upon it and propelled through it at high speed. For full account (very full) of these experiments, and of a later series leading up to the construction of a series of 'model aerodromes' capable of flight under power, it is necessary to turn to the bulky memoir of Smithsonian origin.

The account of these experiments as given by Langley himself reveals the humility of the true investigator. Concerning them, Langley remarks that, 'Everything here has been done with a view to putting a trial aerodrome successfully in flight within a few years, and thus giving an early demonstration of the only kind which is conclusive in the eyes of the scientific man, as well as of the general public--a demonstration that mechanical flight is possible--by actually flying. All that has been done has been with an eye principally to this immediate result, and all the experiments given in this book are to be considered only as approximations to exact truth. All were made with a view, not to some remote future, but to an arrival within the compass of a few years at some result in actual flight that could not be gainsaid or mistaken.'

With a series of over thirty rubber-driven models Langley demonstrated the practicability of opposing curved surfaces to the resistance of the air in such a way as to achieve flight, in the early nineties of last century; he then set about finding the motive power which should permit of the construction of larger machines, up to man-carrying size. The internal combustion engine was then an unknown quantity, and he had to turn to steam, finally, as the propulsive energy for his power plant. The chief problem which faced him was that of the relative weight and power of his engine; he harked back to the Stringfellow engine of 1868, which in 1889 came into the possession of the Smithsonian Institution as a historical curiosity. Rightly or wrongly Langley concluded on examination that this engine never had developed and never could develop more than a tenth of the power attributed to it; consequently he abandoned the idea of copying the Stringfellow design and set about making his own engine.

How he overcame the various difficulties that faced him and constructed a steam-engine capable of the task allotted to it forms a story in itself, too long for recital here. His first power-driven aerodrome of model size was begun in November of 1891, the scale of construction being decided with the idea that it should be large enough to carry an automatic steering apparatus which would render the machine capable of maintaining a long and steady flight. The actual weight of the first model far exceeded the theoretical estimate, and Langley found that a constant increase of weight under the exigencies of construction was a feature which could never be altogether eliminated. The machine was made principally of steel, the sustaining surfaces being composed of silk stretched from a steel tube with wooden attachments. The first engines were the oscillating type, but were found deficient in power. This led to the construction of single-acting inverted oscillating engines with high and low pressure cylinders, and with admission and exhaust ports to avoid the complication and weight of eccentric and valves. Boiler and furnace had to be specially designed; an analysis of sustaining surfaces and the settlement of equilibrium while in flight had to be overcome, and then it was possible to set about the construction of the series of model aerodromes and make test of their 'lift.'

By the time Langley had advanced sufficiently far to consider it possible to conduct experiments in the open air, even with these models, he had got to his fifth aerodrome, and to the year 1894. Certain tests resulted in failure, which in turn resulted in further modifications of design, mainly of the engines. By February of 1895 Langley reported that under favourable conditions a lift of nearly sixty per cent of the flying weight was secured, but although this was much more than was required for flight, it was decided to postpone trials until two machines were ready for the test. May, 1896, came before actual trials were made, when one machine proved successful and another, a later design, failed. The difficulty with these models was that of securing a correct angle for launching; Langley records how, on launching one machine, it rose so rapidly that it attained an angle of sixty degrees and then did a tail slide into the water with its engines working at full speed, after advancing nearly forty feet and remaining in the air for about three seconds. Here, Langley found that he had to obtain greater rigidity in his wings, owing to the distortion of the form of wing under pressure, and how he overcame this difficulty constitutes yet another story too long for the telling here.

Field trials were first attempted in 1893, and Langley blamed his launching apparatus for their total failure. There was a brief, but at the same time practical, success in model flight in 1894, extending to between six and seven seconds, but this only proved the need for strengthening of the wing. In 1895 there was practically no advance toward the solution of the problem, but the flights of May 6th and November 28th, 1896, were notably successful. A diagram given in Langley's memoir shows the track covered by the aerodrome on these two flights; in the first of them the machine made three complete circles, covering a distance of 3,200 feet; in the second, that of November 28th, the distance covered was 4,200 feet, or about three-quarters of a mile, at a speed of about thirty miles an hour.

These achievements meant a good deal; they proved mechanically propelled flight possible. The difference between them and such experiments as were conducted by Clement Ader, Maxim, and others, lay principally in the fact that these latter either did or did not succeed in rising into the air once, and then, either willingly or by compulsion, gave up the quest, while Langley repeated his experiments and thus attained to actual proof of the possibilities of flight. Like these others, however, he decided in 1896 that he would not undertake the construction of a large man-carrying machine. In addition to a multitude of actual duties, which left him practically no time available for original research, he had as an adverse factor fully ten years of disheartening difficulties in connection with his model machines. It was President McKinley who, by requesting Langley to undertake the construction and test of a machine which might finally lead to the development of a flying machine capable of being used in warfare, egged him on to his final experiment. Langley's acceptance of the offer to construct such a machine is contained in a letter addressed from the Smithsonian Institution on December 12th, 1898, to the Board of Ordnance and Fortification of the United States War Department; this letter is of such interest as to render it worthy of reproduction:--

'Gentlemen,--In response to your invitation I repeat what I had the honour to say to the Board--that I am willing, with the consent of the Regents of this Institution, to undertake for the Government the further investigation of the subject of the construction of a flying machine on a scale capable of carrying a man, the investigation to include the construction, development and test of such a machine under conditions left as far as practicable in my discretion, it being understood that my services are given to the Government in such time as may not be occupied by the business of the Institution, and without charge.

'I have reason to believe that the cost of the construction will come within the sum of $50,000.00, and that not more than one-half of that will be called for in the coming year.

'I entirely agree with what I understand to be the wish of the Board that privacy be observed with regard to the work, and only when it reaches a successful completion shall I wish to make public the fact of its success.

'I attach to this a memorandum of my understanding of some points of detail in order to be sure that it is also the understanding of the Board, and I am, gentlemen, with much respect, your obedient servant, S. P. Langley.'

One of the chief problems in connection with the construction of a full-sized apparatus was that of the construction of an engine, for it was realised from the first that a steam power plant for a full-sized machine could only be constructed in such a way as to make it a constant menace to the machine which it was to propel. By this time (1898) the internal combustion engine had so far advanced as to convince Langley that it formed the best power plant available. A contract was made for the delivery of a twelve horse-power engine to weigh not more than a hundred pounds, but this contract was never completed, and it fell to Charles M. Manly to design the five-cylinder radial engine, of which a brief account is included in the section of this work devoted to aero engines, as the power plant for the Langley machine.

The history of the years 1899 to 1903 in the Langley series of experiments contains a multitude of detail far beyond the scope of this present study, and of interest mainly to the designer. There were frames, engines, and propellers, to be considered, worked out, and constructed. We are concerned here mainly with the completed machine and its trials. Of these latter it must be remarked that the only two actual field trials which took place resulted in accidents due to the failure of the launching apparatus, and not due to any inherent defect in the machine. It was intended that these two trials should be the first of a series, but the unfortunate accidents, and the fact that no further funds were forthcoming for continuance of experiments, prevented Langley's success, which, had he been free to go through as he intended with his work, would have been certain.

The best brief description of the Langley aerodrome in its final form, and of the two attempted trials, is contained in the official report of Major M. M. Macomb of the United States Artillery Corps, which report is here given in full:--

REPORT

Experiments with working models which were concluded August 8 last having proved the principles and calculations on which the design of the Langley aerodrome was based to be correct, the next step was to apply these principles to the construction of a machine of sufficient size and power to permit the carrying of a man, who could control the motive power and guide its flight, thus pointing the way to attaining the final goal of producing a machine capable of such extensive and precise aerial flight, under normal atmospheric conditions, as to prove of military or commercial utility.

Mr C. M. Manly, working under Professor Langley, had, by the summer of 1903, succeeded in completing an engine-driven machine which under favourable atmospheric conditions was expected to carry a man for any time up to half an hour, and to be capable of having its flight directed and controlled by him.

The supporting surface of the wings was ample, and experiment showed the engine capable of supplying more than the necessary motive power.

Owing to the necessity of lightness, the weight of the various elements had to be kept at a minimum, and the factor of safety in construction was therefore exceedingly small, so that the machine as a whole was delicate and frail and incapable of sustaining any unusual strain. This defect was to be corrected in later models by utilising data gathered in future experiments under varied conditions.

One of the most remarkable results attained was the production of a gasoline engine furnishing over fifty continuous horse-power for a weight of 120 lbs.

The aerodrome, as completed and prepared for test, is briefly described by Professor Langley as 'built of steel, weighing complete about 730 lbs., supported by 1,040 feet of sustaining surface, having two propellers driven by a gas engine developing continuously over fifty brake horse-power.'

The appearance of the machine prepared for flight was exceedingly light and graceful, giving an impression to all observers of being capable of successful flight.

On October 7 last everything was in readiness, and I witnessed the attempted trial on that day at Widewater, Va. On the Potomac. The engine worked well and the machine was launched at about 12.15 p.m. The trial was unsuccessful because the front guy-post caught in its support on the launching car and was not released in time to give free flight, as was intended, but, on the contrary, caused the front of the machine to be dragged downward, bending the guy-post and making the machine plunge into the water about fifty yards in front of the house-boat. The machine was subsequently recovered and brought back to the house-boat. The engine was uninjured and the frame only slightly damaged, but the four wings and rudder were practically destroyed by the first plunge and subsequent towing back to the house-boat.

This accident necessitated the removal of the house-boat to Washington for the more convenient repair of damages.

On December 8 last, between 4 and 5 p.m., another attempt at a trial was made, this time at the junction of the Anacostia with the Potomac, just below Washington Barracks.

On this occasion General Randolph and myself represented the Board of Ordnance and Fortification. The launching car was released at 4.45 p.m. being pointed up the Anacostia towards the Navy Yard. My position was on the tug Bartholdi, about 150 feet from and at right angles to the direction of proposed flight. The car was set in motion and the propellers revolved rapidly, the engine working perfectly, but there was something wrong with the launching. The rear guy-post seemed to drag, bringing the rudder down on the launching ways, and a crashing, rending sound, followed by the collapse of the rear wings, showed that the machine had been wrecked in the launching, just how, it was impossible for me to see. The fact remains that the rear wings and rudder were wrecked before the machine was free of the ways. Their collapse deprived the machine of its support in the rear, and it consequently reared up in front under the action of the motor, assumed a vertical position, and then toppled over to the rear, falling into the water a few feet in front of the boat.

Mr Manly was pulled out of the wreck uninjured and the wrecked machine--was subsequently placed upon the house-boat, and the whole brought back to Washington.

From what has been said it will be seen that these unfortunate accidents have prevented any test of the apparatus in free flight, and the claim that an engine-driven, man-carrying aerodrome has been constructed lacks the proof which actual flight alone can give.

Having reached the present stage of advancement in its development, it would seem highly desirable, before laying down the investigation, to obtain conclusive proof of the possibility of free flight, not only because there are excellent reasons to hope for success, but because it marks the end of a definite step toward the attainment of the final goal.

Just what further procedure is necessary to secure successful flight with the large aerodrome has not yet been decided upon. Professor Langley is understood to have this subject under advisement, and will doubtless inform the Board of his final conclusions as soon as practicable.

In the meantime, to avoid any possible misunderstanding, it should be stated that even after a successful test of the present great aerodrome, designed to carry a man, we are still far from the ultimate goal, and it would seem as if years of constant work and study by experts, together with the expenditure of thousands of dollars, would still be necessary before we can hope to produce an apparatus of practical utility on these lines.--Washington, January 6, 1904.

A subsequent report of the Board of ordnance and Fortification to the Secretary of War embodied the principal points in Major Macomb's report, but as early as March 3rd, 1904, the Board came to a similar conclusion to that of the French Ministry of War in respect of Clement Ader's work, stating that it was not 'prepared to make an additional allotment at this time for continuing the work.' This decision was in no small measure due to hostile newspaper criticisms. Langley, in a letter to the press explaining his attitude, stated that he did not wish to make public the results of his work till these were certain, in consequence of which he refused admittance to newspaper representatives, and this attitude produced a hostility which had effect on the United States Congress. An offer was made to commercialise the invention, but Langley steadfastly refused it. Concerning this, Manly remarks that Langley had 'given his time and his best labours to the world without hope of remuneration, and he could not bring himself, at his stage of life, to consent to capitalise his scientific work.'

The final trial of the Langley aerodrome was made on December 8th, 1903; nine days later, on December 17th, the Wright Brothers made their first flight in a power-propelled machine, and the conquest of the air was thus achieved. But for the two accidents that spoilt his trials, the honour which fell to the Wright Brothers would, beyond doubt, have been secured by Samuel Pierpoint Langley.

XI. THE WRIGHT BROTHERS

Such information as is given here concerning the Wright Brothers is derived from the two best sources available, namely, the writings of Wilbur Wright himself, and a lecture given by Dr Griffith Brewer to members of the Royal Aeronautical Society. There is no doubt that so far as actual work in connection with aviation accomplished by the two brothers is concerned, Wilbur Wright's own statements are the clearest and best available. Apparently Wilbur was, from the beginning, the historian of the pair, though he himself would have been the last to attempt to detract in any way from the fame that his brother's work also deserves. Throughout all their experiments the two were inseparable, and their work is one indivisible whole; in fact, in every department of that work, it is impossible to say where Orville leaves off and where Wilbur begins.

It is a great story, this of the Wright Brothers, and one worth all the detail that can be spared it. It begins on the 16th April, 1867, when Wilbur Wright was born within eight miles of Newcastle, Indiana. Before Orville's birth on the 19th August, 1871, the Wright family had moved to Dayton, Ohio, and settled on what is known as the 'West Side' of the town. Here the brothers grew up, and, when Orville was still a boy in his teens, he started a printing business, which, as Griffith Brewer remarks, was only limited by the smallness of his machine and small quantity of type at his disposal. This machine was in such a state that pieces of string and wood were incorporated in it by way of repair, but on it Orville managed to print a boys' paper which gained considerable popularity in Dayton 'West Side.' Later, at the age of seventeen, he obtained a more efficient outfit, with which he launched a weekly newspaper, four pages in size, entitled The West Side News. After three months' running the paper was increased in size and Wilbur came into the enterprise as editor, Orville remaining publisher. In 1894 the two brothers began the publication of a weekly magazine, Snap-Shots, to which Wilbur contributed a series of articles on local affairs that gave evidence of the incisive and often sarcastic manner in which he was able to express himself throughout his life. Dr Griffith Brewer describes him as a fearless critic, who wrote on matters of local interest in a kindly but vigorous manner, which did much to maintain the healthy public municipal life of Dayton.

Editorial and publishing enterprise was succeeded by the formation, just across the road from the printing works, of the Wright Cycle Company, where the two brothers launched out as cycle manufacturers with the 'Van Cleve' bicycle, a machine of great local repute for excellence of construction, and one which won for itself a reputation that lasted long after it had ceased to be manufactured. The name of the machine was that of an ancestor of the brothers, Catherine Van Cleve, who was one of the first settlers at Dayton, landing there from the River Miami on April 1st, 1796, when the country was virgin forest.

It was not until 1896 that the mechanical genius which characterised the two brothers was turned to the consideration of aeronautics. In that year they took up the problem thoroughly, studying all the aeronautical information then in print. Lilienthal's writings formed one basis for their studies, and the work of Langley assisted in establishing in them a confidence in the possibility of a solution to the problems of mechanical flight. In 1909, at the banquet given by the Royal Aero Club to the Wright Brothers on their return to America, after the series of demonstration flights carried out by Wilbur Wright on the Continent, Wilbur paid tribute to the great pioneer work of Stringfellow, whose studies and achievements influenced his own and Orville's early work. He pointed out how Stringfellow devised an aeroplane having two propellers and vertical and horizontal steering, and gave due place to this early pioneer of mechanical flight.

Neither of the brothers was content with mere study of the work of others. They collected all the theory available in the books published up to that time, and then built man-carrying gliders with which to test the data of Lilienthal and such other authorities as they had consulted. For two years they conducted outdoor experiments in order to test the truth or otherwise of what were enunciated as the principles of flight; after this they turned to laboratory experiments, constructing a wind tunnel in which they made thousands of tests with models of various forms of curved planes. From their experiments they tabulated thousands of readings, which Griffith Brewer remarks as giving results equally efficient with those of the elaborate tables prepared by learned institutions.

Wilbur Wright has set down the beginnings of the practical experiments made by the two brothers very clearly. 'The difficulties,' he says, 'which obstruct the pathway to success in flying machine construction are of three general classes: (1) Those which relate to the construction of the sustaining wings; (2) those which relate to the generation and application of the power required to drive the machine through the air; (3) those relating to the balancing and steering of the machine after it is actually in flight. Of these difficulties two are already to a certain extent solved. Men already know how to construct wings, or aeroplanes, which, when driven through the air at sufficient speed, will not only sustain the weight of the wings themselves, but also that of the engine and the engineer as well. Men also know how to build engines and' screws of sufficient lightness and power to drive these planes at sustaining speed. Inability to balance and steer still confronts students of the flying problem, although nearly ten years have passed (since Lilienthal's success). When this one feature has been worked out, the age of flying machines will have arrived, for all other difficulties are of minor importance.

'The person who merely watches the flight of a bird gathers the impression that the bird has nothing to think of but the flapping of its wings. As a matter of fact, this is a very small part of its mental labour. Even to mention all the things the bird must constantly keep in mind in order to fly securely through the air would take a considerable time. If I take a piece of paper and, after placing it parallel with the ground, quickly let it fall, it will not settle steadily down as a staid, sensible piece of paper ought to do, but it insists on contravening every recognised rule of decorum, turning over and darting hither and thither in the most erratic manner, much after the style of an untrained horse. Yet this is the style of steed that men must learn to manage before flying can become an everyday sport. The bird has learned this art of equilibrium, and learned it so thoroughly that its skill is not apparent to our sight. We only learn to appreciate it when we can imitate it.

'Now, there are only two ways of learning to ride a fractious horse: one is to get on him and learn by actual practice how each motion and trick may be best met; the other is to sit on a fence and watch the beast awhile, and then retire to the house and at leisure figure out the best way of overcoming his jumps and kicks. The latter system is the safer, but the former, on the whole, turns out the larger proportion of good riders. It is very much the same in learning to ride a flying machine; if you are looking for perfect safety you will do well to sit on a fence and watch the birds, but if you really wish to learn you must mount a machine and become acquainted with its tricks by actual trial. The balancing of a gliding or flying machine is very simple in theory. It merely consists in causing the centre of pressure to coincide with the centre of gravity.'

These comments are taken from a lecture delivered by Wilbur Wright before the Western Society of Engineers in September of 1901, under the presidency of Octave Chanute. In that lecture Wilbur detailed the way in which he and his brother came to interest themselves in aeronautical problems and constructed their first glider. He speaks of his own notice of the death of Lilienthal in 1896, and of the way in which this fatality roused him to an active interest in aeronautical problems, which was stimulated by reading Professor Marey's Animal Mechanism, not for the first time. 'From this I was led to read more modern works, and as my brother soon became equally interested with myself, we soon passed from the reading to the thinking, and finally to the working stage. It seemed to us that the main reason why the problem had remained so long unsolved was that no one had been able to obtain any adequate practice. We figured that Lilienthal in five years of time had spent only about five hours in actual gliding through the air. The wonder was not that he had done so little, but that he had accomplished so much. It would not be considered at all safe for a bicycle rider to attempt to ride through a crowded city street after only five hours' practice, spread out in bits of ten seconds each over a period of five years; yet Lilienthal with this brief practice was remarkably successful in meeting the fluctuations and eddies of wind-gusts. We thought that if some method could be found by which it would be possible to practice by the hour instead of by the second there would be hope of advancing the solution of a very difficult problem. It seemed feasible to do this by building a machine which would be sustained at a speed of eighteen miles per hour, and then finding a locality where winds of this velocity were common. With these conditions a rope attached to the machine to keep it from floating backward would answer very nearly the same purpose as a propeller driven by a motor, and it would be possible to practice by the hour, and without any serious danger, as it would not be necessary to rise far from the ground, and the machine would not have any forward motion at all. We found, according to the accepted tables of air pressure on curved surfaces, that a machine spreading 200 square feet of wing surface would be sufficient for our purpose, and that places would easily be found along the Atlantic coast where winds of sixteen to twenty-five miles were not at all uncommon. When the winds were low it was our plan to glide from the tops of sandhills, and when they were sufficiently strong to use a rope for our motor and fly over one spot. Our next work was to draw up the plans for a suitable machine. After much study we finally concluded that tails were a source of trouble rather than of assistance, and therefore we decided to dispense with them altogether. It seemed reasonable that if the body of the operator could be placed in a horizontal position instead of the upright, as in the machines of Lilienthal, Pilcher, and Chanute, the wind resistance could be very materially reduced, since only one square foot instead of five would be exposed. As a full half horse-power would be saved by this change, we arranged to try at least the horizontal position. Then the method of control used by Lilienthal, which consisted in shifting the body, did not seem quite as quick or effective as the case required; so, after long study, we contrived a system consisting of two large surfaces on the Chanute double-deck plan, and a smaller surface placed a short distance in front of the main surfaces in such a position that the action of the wind upon it would counterbalance the effect of the travel of the centre of pressure on the main surfaces. Thus changes in the direction and velocity of the wind would have little disturbing effect, and the operator would be required to attend only to the steering of the machine, which was to be effected by curving the forward surface up or down. The lateral equilibrium and the steering to right or left was to be attained by a peculiar torsion of the main surfaces which was equivalent to presenting one end of the wings at a greater angle than the other. In the main frame a few changes were also made in the details of construction and trussing employed by Mr Chanute. The most important of these were: (1) The moving of the forward main crosspiece of the frame to the extreme front edge; (2) the encasing in the cloth of all crosspieces and ribs of the surfaces; (3) a rearrangement of the wires used in trussing the two surfaces together, which rendered it possible to tighten all the wires by simply shortening two of them.'

The brothers intended originally to get 200 square feet of supporting surface for their glider, but the impossibility of obtaining suitable material compelled them to reduce the area to 165 square feet, which, by the Lilienthal tables, admitted of support in a wind of about twenty-one miles an hour at an angle of three degrees. With this glider they went in the summer of I 1900 to the little settlement of Kitty Hawk, North Carolina, situated on the strip of land dividing Albemarle Sound from the Atlantic. Here they reckoned on obtaining steady wind, and here, on the day that they completed the machine, they took it out for trial as a kite with the wind blowing at between twenty-five and thirty miles an hour. They found that in order to support a man on it the glider required an angle nearer twenty degrees than three, and even with the wind at thirty miles an hour they could not get down to the planned angle of three degrees. 'Later, when the wind was too light to support the machine with a man on it, they tested it as a kite, working the rudders by cords. Although they obtained satisfactory results in this way they realised fully that actual gliding experience was necessary before the tests could be considered practical.

A series of actual measurements of lift and drift of the machine gave astonishing results. 'It appeared that the total horizontal pull of the machine, while sustaining a weight of 52 lbs., was only 8.5 lbs., which was less than had been previously estimated for head resistance of the framing alone. Making allowance for the weight carried, it appeared that the head resistance of the framing was but little more than fifty per cent of the amount which Mr Chanute had estimated as the head resistance of the framing of his machine. On the other hand, it appeared sadly deficient in lifting power as compared with the calculated lift of curved surfaces of its size... we decided to arrange our machine for the following year so that the depth of curvature of its surfaces could be varied at will, and its covering air-proofed.'

After these experiments the brothers decided to turn to practical gliding, for which they moved four miles to the south, to the Kill Devil sandhills, the principal of which is slightly over a hundred feet in height, with an inclination of nearly ten degrees on its main north-western slope. On the day after their arrival they made about a dozen glides, in which, although the landings were made at a speed of more than twenty miles an hour, no injury was sustained either by the machine or by the operator.

'The slope of the hill was 9.5 degrees, or a drop of one foot in six. We found that after attaining a speed of about twenty-five to thirty miles with reference to the wind, or ten to fifteen miles over the ground, the machine not only glided parallel to the slope of the hill, but greatly increased its speed, thus indicating its ability to glide on a somewhat less angle than 9.5 degrees, when we should feel it safe to rise higher from the surface. The control of the machine proved even better than we had dared to expect, responding quickly to the slightest motion of the rudder. With these glides our experiments for the year 1900 closed. Although the hours and hours of practice we had hoped to obtain finally dwindled down to about two minutes, we were very much pleased with the general results of the trip, for, setting out as we did with almost revolutionary theories on many points and an entirely untried form of machine, we considered it quite a point to be able to return without having our pet theories completely knocked on the head by the hard logic of experience, and our own brains dashed out in the bargain. Everything seemed to us to confirm the correctness of our original opinions: (1) That practice is the key to the secret of flying; (2) that it is practicable to assume the horizontal position; (3) that a smaller surface set at a negative angle in front of the main bearing surfaces, or wings, will largely counteract the effect of the fore and aft travel of the centre of pressure; (4) that steering up and down can be attained with a rudder without moving the position of the operator's body; (5) that twisting the wings so as to present their ends to the wind at different angles is a more prompt and efficient way of maintaining lateral equilibrium than shifting the body of the operator.'

For the gliding experiments of 1901 it was decided to retain the form of the 1900 glider, but to increase the area to 308 square feet, which, the brothers calculated, would support itself and its operator in a wind of seventeen miles an hour with an angle of incidence of three degrees. Camp was formed at Kitty Hawk in the middle of July, and on July 27th the machine was completed and tried for the first time in a wind of about fourteen miles an hour. The first attempt resulted in landing after a glide of only a few yards, indicating that the centre of gravity was too far in front of the centre of pressure. By shifting his position farther and farther back the operator finally achieved an undulating flight of a little over 300 feet, but to obtain this success he had to use full power of the rudder to prevent both stalling and nose-diving. With the 1900 machine one-fourth of the rudder action had been necessary for far better control.

Practically all glides gave the same result, and in one the machine rose higher and higher until it lost all headway. 'This was the position from which Lilienthal had always found difficulty in extricating himself, as his machine then, in spite of his greatest exertions, manifested a tendency to dive downward almost vertically and strike the ground head on with frightful velocity. In this case a warning cry from the ground caused the operator to turn the rudder to its full extent and also to move his body slightly forward. The machine then settled slowly to the ground, maintaining its horizontal position almost perfectly, and landed without any injury at all. This was very encouraging, as it showed that one of the very greatest dangers in machines with horizontal tails had been overcome by the use of the front rudder. Several glides later the same experience was repeated with the same result. In the latter case the machine had even commenced to move backward, but was nevertheless brought safely to the ground in a horizontal position. On the whole this day's experiments were encouraging, for while the action of the rudder did not seem at all like that of our 1900 machine, yet we had escaped without difficulty from positions which had proved very dangerous to preceding experimenters, and after less than one minute's actual practice had made a glide of more than 300 feet, at an angle of descent of ten degrees, and with a machine nearly twice as large as had previously been considered safe. The trouble with its control, which has been mentioned, we believed could be corrected when we should have located its cause.'

It was finally ascertained that the defect could be remedied by trussing down the ribs of the whole machine so as to reduce the depth of curvature. When this had been done gliding was resumed, and after a few trials glides of 366 and 389 feet were made with prompt response on the part of the machine, even to small movements of the rudder. The rest of the story of the gliding experiments of 1901 cannot be better told than in Wilbur Wright's own words, as uttered by him in the lecture from which the foregoing excerpts have been made.

'The machine, with its new curvature, never failed to respond promptly to even small movements of the rudder. The operator could cause it to almost skim the ground, following the undulations of its surface, or he could cause it to sail out almost on a level with the starting point, and, passing high above the foot of the hill, gradually settle down to the ground. The wind on this day was blowing eleven to fourteen miles per hour. The next day, the conditions being favourable, the machine was again taken out for trial. This time the velocity of the wind was eighteen to twenty-two miles per hour. At first we felt some doubt as to the safety of attempting free flight in so strong a wind, with a machine of over 300 square feet and a practice of less than five minutes spent in actual flight. But after several preliminary experiments we decided to try a glide. The control of the machine seemed so good that we then felt no apprehension in sailing boldly forth. And thereafter we made glide after glide, sometimes following the ground closely and sometimes sailing high in the air. Mr Chanute had his camera with him and took pictures of some of these glides, several of which are among those shown.

'We made glides on subsequent days, whenever the conditions were favourable. The highest wind thus experimented in was a little over twelve metres per second--nearly twenty-seven miles per hour.

It had been our intention when building the machine to do the larger part of the experimenting in the following manner:--When the wind blew seventeen miles an hour, or more, we would attach a rope to the machine and let it rise as a kite with the operator upon it. When it should reach a proper height the operator would cast off the rope and glide down to the ground just as from the top of a hill. In this way we would be saved the trouble of carrying the machine uphill after each glide, and could make at least ten glides in the time required for one in the other way. But when we came to try it, we found that a wind of seventeen miles, as measured by Richards' anemometer, instead of sustaining the machine with its operator, a total weight of 240 lbs., at an angle of incidence of three degrees, in reality would not sustain the machine alone--100 lbs.--at this angle. Its lifting capacity seemed scarcely one third of the calculated amount. In order to make sure that this was not due to the porosity of the cloth, we constructed two small experimental surfaces of equal size, one of which was air-proofed and the other left in its natural state; but we could detect no difference in their lifting powers. For a time we were led to suspect that the lift of curved surfaces very little exceeded that of planes of the same size, but further investigation and experiment led to the opinion that (1) the anemometer used by us over-recorded the true velocity of the wind by nearly 15 per cent; (2) that the well-known Smeaton co-efficient of .005 V squared for the wind pressure at 90 degrees is probably too great by at least 20 per cent; (3) that Lilienthal's estimate that the pressure on a curved surface having an angle of incidence of 3 degrees equals .545 of the pressure at go degrees is too large, being nearly 50 per cent greater than very recent experiments of our own with a pressure testing-machine indicate; (4) that the superposition of the surfaces somewhat reduced the lift per square foot, as compared with a single surface of equal area.

'In gliding experiments, however, the amount of lift is of less relative importance than the ratio of lift to drift, as this alone decides the angle of gliding descent.

In a plane the pressure is always perpendicular to the surface, and the ratio of lift to drift is therefore the same as that of the cosine to the sine of the angle of incidence. But in curved surfaces a very remarkable situation is found. The pressure, instead of being uniformly normal to the chord of the arc, is usually inclined considerably in front of the perpendicular. The result is that the lift is greater and the drift less than if the pressure were normal. Lilienthal was the first to discover this exceedingly important fact, which is fully set forth in his book, Bird Flight the Basis of the Flying Art, but owing to some errors in the methods he used in making measurements, question was raised by other investigators not only as to the accuracy of his figures, but even as to the existence of any tangential force at all. Our experiments confirm the existence of this force, though our measurements differ considerably from those of Lilienthal. While at Kitty Hawk we spent much time in measuring the horizontal pressure on our unloaded machine at various angles of incidence. We found that at 13 degrees the horizontal pressure was about 23 lbs. This included not only the drift proper, or horizontal component of the pressure on the side of the surface, but also the head resistance of the framing as well. The weight of the machine at the time of this test was about 108 lbs. Now, if the pressure had been normal to the chord of the surface, the drift proper would have been to the lift (108 lbs.) as the sine of 13 degrees is to the cosine of 13 degrees, or .22 X 108/.97 = 24+ lbs.; but this slightly exceeds the total pull of 23 pounds on our scales. Therefore it is evident that the average pressure on the surface, instead of being normal to the chord, was so far inclined toward the front that all the head resistance of framing and wires used in the construction was more than overcome. In a wind of fourteen miles per hour resistance is by no means a negligible factor, so that tangential is evidently a force of considerable value. In a higher wind, which sustained the machine at an angle of 10 degrees the pull on the scales was 18 lbs. With the pressure normal to the chord the drift proper would have been 17 X 98/.98. The travel of the centre of pressure made it necessary to put sand on the front rudder to bring the centres of gravity and pressure into coincidence, consequently the weight of the machine varied from 98 lbs. to 108 lbs. in the different tests= 17 lbs., so that, although the higher wind velocity must have caused an increase in the head resistance, the tangential force still came within 1 lb. of overcoming it. After our return from Kitty Hawk we began a series of experiments to accurately determine the amount and direction of the pressure produced on curved surfaces when acted upon by winds at the various angles from zero to 90 degrees. These experiments are not yet concluded, but in general they support Lilienthal in the claim that the curves give pressures more favourable in amount and direction than planes; but we find marked differences in the exact values, especially at angles below 10 degrees. We were unable to obtain direct measurements of the horizontal pressures of the machine with the operator on board, but by comparing the distance travelled with the vertical fall, it was easily calculated that at a speed of 24 miles per hour the total horizontal resistances of our machine, when bearing the operator, amounted to 40 lbs., which is equivalent to about 2 1/3 horse-power. It must not be supposed, however, that a motor developing this power would be sufficient to drive a man-bearing machine. The extra weight of the motor would require either a larger machine, higher speed, or a greater angle of incidence in order to support it, and therefore more power. It is probable, however, that an engine of 6 horse-power, weighing 100 lbs. would answer the purpose. Such an engine is entirely practicable. Indeed, working motors of one-half this weight per horse-power (9 lbs. per horse-power) have been constructed by several different builders. Increasing the speed of our machine from 24 to 33 miles per hour reduced the total horizontal pressure from 40 to about 35 lbs. This was quite an advantage in gliding, as it made it possible to sail about 15 per cent farther with a given drop. However, it would be of little or no advantage in reducing the size of the motor in a power-driven machine, because the lessened thrust would be counterbalanced by the increased speed per minute. Some years ago Professor Langley called attention to the great economy of thrust which might be obtained by using very high speeds, and from this many were led to suppose that high speed was essential to success in a motor-driven machine. But the economy to which Professor Langley called attention was in foot pounds per mile of travel, not in foot pounds per minute. It is the foot pounds per minute that fixes the size of the motor. The probability is that the first flying machines will have a relatively low speed, perhaps not much exceeding 20 miles per hour, but the problem of increasing the speed will be much simpler in some respects than that of increasing the speed of a steamboat; for, whereas in the latter case the size of the engine must increase as the cube of the speed, in the flying machine, until extremely high speeds are reached, the capacity of the motor increases in less than simple ratio; and there is even a decrease in the fuel per mile of travel. In other words, to double the speed of a steamship (and the same is true of the balloon type of airship) eight times the engine and boiler capacity would be required, and four times the fuel consumption per mile of travel: while a flying machine would require engines of less than double the size, and there would be an actual decrease in the fuel consumption per mile of travel. But looking at the matter conversely, the great disadvantage of the flying machine is apparent; for in the latter no flight at all is possible unless the proportion of horse-power to flying capacity is very high; but on the other hand a steamship is a mechanical success if its ratio of horse-power to tonnage is insignificant. A flying machine that would fly at a speed of 50 miles per hour with engines of 1,000 horse-power would not be upheld by its wings at all at a speed of less than 25 miles an hour, and nothing less than 500 horse-power could drive it at this speed. But a boat which could make 40 miles an hour with engines of 1,000 horse-power would still move 4 miles an hour even if the engines were reduced to 1 horse-power. The problems of land and water travel were solved in the nineteenth century, because it was possible to begin with small achievements, and gradually work up to our present success. The flying problem was left over to the twentieth century, because in this case the art must be highly developed before any flight of any considerable duration at all can be obtained.

'However, there is another way of flying which requires no artificial motor, and many workers believe that success will come first by this road. I refer to the soaring flight, by which the machine is permanently sustained in the air by the same means that are employed by soaring birds. They spread their wings to the wind, and sail by the hour, with no perceptible exertion beyond that required to balance and steer themselves. What sustains them is not definitely known, though it is almost certain that it is a rising current of air. But whether it be a rising current or something else, it is as well able to support a flying machine as a bird, if man once learns the art of utilising it. In gliding experiments it has long been known that the rate of vertical descent is very much retarded, and the duration of the flight greatly prolonged, if a strong wind blows UP the face of the hill parallel to its surface. Our machine, when gliding in still air, has a rate of vertical descent of nearly 6 feet per second, while in a wind blowing 26 miles per hour up a steep hill we made glides in which the rate of descent was less than 2 feet per second. And during the larger part of this time, while the machine remained exactly in the rising current, THERE WAS NO

DESCENT AT ALL, BUT EVEN A SLIGHT RISE. If the operator had had sufficient skill to keep himself from passing beyond the rising current he would have been sustained indefinitely at a higher point than that from which he started. The illustration shows one of these very slow glides at a time when the machine was practically at a standstill. The failure to advance more rapidly caused the photographer some trouble in aiming, as you will perceive. In looking at this picture you will readily understand that the excitement of gliding experiments does not entirely cease with the breaking up of camp. In the photographic dark-room at home we pass moments of as thrilling interest as any in the field, when the image begins to appear on the plate and it is yet an open question whether we have a picture of a flying machine or merely a patch of open sky. These slow glides in rising current probably hold out greater hope of extensive practice than any other method within man's reach, but they have the disadvantage of requiring rather strong winds or very large supporting surfaces. However, when gliding operators have attained greater skill, they can with comparative safety maintain themselves in the air for hours at a time in this way, and thus by constant practice so increase their knowledge and skill that they can rise into the higher air and search out the currents which enable the soaring birds to transport themselves to any desired point by first rising in a circle and then sailing off at a descending angle. This illustration shows the machine, alone, flying in a wind of 35 miles per hour on the face of a steep hill, 100 feet high. It will be seen that the machine not only pulls upward, but also pulls forward in the direction from which the wind blows, thus overcoming both gravity and the speed of the wind. We tried the same experiment with a man on it, but found danger that the forward pull would become so strong, that the men holding the ropes would be dragged from their insecure foothold on the slope of the hill. So this form of experimenting was discontinued after four or five minutes' trial.

'In looking over our experiments of the past two years, with models and full-size machines, the following points stand out with clearness:--

'1. That the lifting power of a large machine, held stationary in a wind at a small distance from the earth, is much less than the Lilienthal table and our own laboratory experiments would lead us to expect. When the machine is moved through the air, as in gliding, the discrepancy seems much less marked.

'2. That the ratio of drift to lift in well-shaped surfaces is less at angles of incidence of 5 degrees to 12 degrees than at an angle of 3 degrees.

'3. That in arched surfaces the centre of pressure at 90 degrees is near the centre of the surface, but moves slowly forward as the angle becomes less, till a critical angle varying with the shape and depth of the curve is reached, after which it moves rapidly toward the rear till the angle of no lift is found.

'4. That with similar conditions large surfaces may be controlled with not much greater difficulty than small ones, if the control is effected by manipulation of the surfaces themselves, rather than by a movement of the body of the operator.

'5. That the head resistances of the framing can be brought to a point much below that usually estimated as necessary.

'6. That tails, both vertical and horizontal, may with safety be eliminated in gliding and other flying experiments.

'7. That a horizontal position of the operator's body may be assumed without excessive danger, and thus the head resistance reduced to about one-fifth that of the upright position.

'8. That a pair of superposed, or tandem surfaces, has less lift in proportion to drift than either surface separately, even after making allowance for weight and head resistance of the connections.'

Thus, to the end of the 1901 experiments, Wilbur Wright provided a fairly full account of what was accomplished; the record shows an amount of patient and painstaking work almost beyond belief--it was no question of making a plane and launching it, but a business of trial and error, investigation and tabulation of detail, and the rejection time after time of previously accepted theories, till the brothers must have felt the the solid earth was no longer secure, at times. Though it was Wilbur who set down this and other records of the work done, yet the actual work was so much Orville's as his brother's that no analysis could separate any set of experiments and say that Orville did this and Wilbur that--the two were inseparable. On this point Griffith Brewer remarked that 'in the arguments, if one brother took one view, the other brother took the opposite view as a matter of course, and the subject was thrashed to pieces until a mutually acceptable result remained. I have often been asked since these pioneer days, "Tell me, Brewer, who was really the originator of those two?" In reply, I used first to say, "I think it was mostly Wilbur," and later, when I came to know Orville better, I said, "The thing could not have been without Orville." Now, when asked, I have to say, "I don't know," and I feel the more I think of it that it was only the wonderful combination of these two brothers, who devoted their lives together or this common object, that made the discovery of the art of flying possible.'

Beyond the 1901 experiments in gliding, the record grows more scrappy, less detailed. It appears that once power-driven flight had been achieved, the brothers were not so willing to talk as before; considering the amount of work that they put in, there could have been little time for verbal description of that work--as already remarked, their tables still stand for the designer and experimenter. The end of the 1901 experiments left both brothers somewhat discouraged, though they had accomplished more than any others. 'Having set out with absolute faith in the existing scientific data, we ere driven to doubt one thing after another, finally, after two years of experiment, we cast it all aside, and decided to rely entirely on our own investigations. Truth and error were everywhere so intimately mixed as to be indistinguishable.... We had taken up aeronautics as a sport. We reluctantly entered upon the scientific side of it.'

Yet, driven thus to the more serious aspect of the work, they found in the step its own reward, for the work of itself drew them on and on, to the construction of measuring machines for the avoidance of error, and to the making of series after series of measurements, concerning which Wilbur wrote in 1908 (in the Century Magazine) that 'after making preliminary measurements on a great number of different shaped surfaces, to secure a general understanding of the subject, we began systematic measurements of standard surfaces, so varied in design as to bring out the underlying causes of differences noted in their pressures. Measurements were tabulated on nearly fifty of these at all angles from zero to 45 degrees, at intervals of 2 1/2 degrees. Measurements were also secured showing the effects on each other when surfaces are superposed, or when they follow one another.

'Some strange results were obtained. One surface, with a heavy roll at the front edge, showed the same lift for all angles from 7 1/2 to 45 degrees. This seemed so anomalous that we were almost ready to doubt our own measurements, when a simple test was suggested. A weather vane, with two planes attached to the pointer at an angle of 80 degrees with each other, was made. According to our table, such a vane would be in unstable equilibrium when pointing directly into the wind, for if by chance the wind should happen to strike one plane at 39 degrees and the other at 41 degrees, the plane with the smaller angle would have the greater pressure and the pointer would be turned still farther out of the course of the wind until the two vanes again secured equal pressures, which would be at approximately 30 and 50 degrees. But the vane performed in this very manner. Further corroboration of the tables was obtained in experiments with the new glider at Kill Devil Hill the next season.

'In September and October, 1902 nearly 1,000 gliding flights were made, several of which covered distances of over 600 feet. Some, made against a wind of 36 miles an hour, gave proof of the effectiveness of the devices for control. With this machine, in the autumn of 1903, we made a number of flights in which we remained in the air for over a minute, often soaring for a considerable time in one spot, without any descent at all. Little wonder that our unscientific assistant should think the only thing needed to keep it indefinitely in the air would be a coat of feathers to make it light!'

It was at the conclusion of these experiments of 1903 that the brothers concluded they had obtained sufficient data from their thousands of glides and multitude of calculations to permit of their constructing and making trial of a power-driven machine. The first designs got out provided for a total weight of 600 lbs., which was to include the weight of the motor and the pilot; but on completion it was found that there was a surplus of power from the motor, and thus they had 150 lbs. weight to allow for strengthening wings and other parts.

They came up against the problem to which Riach has since devoted so much attention, that of propeller design. 'We had thought of getting the theory of the screw-propeller from the marine engineers, and then, by applying our table of air-pressures to their formulae, of designing air-propellers suitable for our uses. But, so far as we could learn, the marine engineers possessed only empirical formulae, and the exact action of the screw propeller, after a century of use, was still very obscure. As we were not in a position to undertake a long series of practical experiments to discover a propeller suitable for our machine, it seemed necessary to obtain such a thorough understanding of the theory of its reactions as would enable us to design them from calculation alone. What at first seemed a simple problem became more complex the longer we studied it. With the machine moving forward, the air flying backward, the propellers turning sidewise, and nothing standing still, it seemed impossible to find a starting point from which to trace the various simultaneous reactions. Contemplation of it was confusing. After long arguments we often found ourselves in the ludicrous position of each having been converted to the other's side, with no more agreement than when the discussion began.

'It was not till several months had passed, and every phase of the problem had been thrashed over and over, that the various reactions began to untangle themselves. When once a clear understanding had been obtained there was no difficulty in designing a suitable propeller, with proper diameter, pitch, and area of blade, to meet the requirements of the flier. High efficiency in a screw-propeller is not dependent upon any particular or peculiar shape, and there is no such thing as a "best" screw. A propeller giving a high dynamic efficiency when used upon one machine may be almost worthless when used upon another. The propeller should in every case be designed to meet the particular conditions of the machine to which it is to be applied. Our first propellers, built entirely from calculation, gave in useful work 66 per cent of the power expended. This was about one-third more than had been secured by Maxim or Langley.'

Langley had made his last attempt with the 'aerodrome,' and his splendid failure but a few days before the brothers made their first attempt at power-driven aeroplane flight. On December 17th, 1903, the machine was taken out; in addition to Wilbur and Orville Wright, there were present five spectators: Mr A. D. Etheridge, of the Kill Devil life-saving station; Mr W. S. Dough, Mr W. C. Brinkley, of Manteo; Mr John Ward, of Naghead, and Mr John T. Daniels[3]. A general invitation had been given to practically all the residents in the vicinity, but the Kill Devil district is a cold area in December, and history had recorded so many experiments in which machines had failed to leave the ground that between temperature and scepticism only these five risked a waste of their time.

And these five were in at the greatest conquest man had made since James Watt evolved the steam engine--perhaps even a greater conquest than that of Watt. Four flights in all were made; the first lasted only twelve seconds, 'the first in the history of the world in which a machine carrying a man had raised itself into the air by its own power in free flight, had sailed forward on a level course without reduction

3 This list is as given by Wilbur Wright himself.

of speed, and had finally landed without being wrecked,' said Wilbur Wright concerning the achievement[4]. The next two flights were slightly longer, and the fourth and last of the day was one second short of the complete minute; it was made into the teeth of a 20 mile an hour wind, and the distance travelled was 852 feet.

This bald statement of the day's doings is as Wilbur Wright himself has given it, and there is in truth nothing more to say; no amount of statement could add to the importance of the achievement, and no more than the bare record is necessary. The faith that had inspired the long roll of pioneers, from da Vinci onward, was justified at last.

Having made their conquest, the brothers took the machine back to camp, and, as they thought, placed it in safety. Talking with the little group of spectators about the flights, they forgot about the machine, and then a sudden gust of wind struck it. Seeing that it was being overturned, all made a rush toward it to save it, and Mr Daniels, a man of large proportions, was in some way lifted off his feet, falling between the planes. The machine overturned fully, and Daniels was shaken like a die in a cup as the wind rolled the machine over and over--he came out at the end of his experience with a series of bad bruises, and no more, but the damage done to the machine by the accident was sufficient to render it useless for further experiment that season.

A new machine, stronger and heavier, was constructed by the brothers, and in the spring of 1904 they began experiments again at Sims Station, eight miles to the east of Dayton, their home town. Press representatives were invited for the first trial, and about a dozen came--the whole gathering did not number more than fifty people. 'When preparations had been concluded,' Wilbur Wright wrote of this trial, 'a wind of only three or four miles an hour was blowing--insufficient for starting on so short a track--but since many had come a long way to see the machine in action, an attempt was made. To add to the other difficulty, the engine refused to work properly. The machine, after running the length of the track, slid off the end without rising into the air at all. Several of the newspaper men returned next day but were again disappointed. The engine performed badly, and after a glide of only sixty feet the machine again came to the ground. Further trial was postponed till the motor could be put in better running condition. The reporters had now, no

---
4 Century Magazine, September, 1908.

doubt, lost confidence in the machine, though their reports, in kindness, concealed it. Later, when they heard that we were making flights of several minutes' duration, knowing that longer flights had been made with airships, and not knowing any essential difference between airships and flying machines, they were but little interested.

'We had not been flying long in 1904 before we found that the problem of equilibrium had not as yet been entirely solved. Sometimes, in making a circle, the machine would turn over sidewise despite anything the operator could do, although, under the same conditions in ordinary straight flight it could have been righted in an instant. In one flight, in 1905, while circling round a honey locust-tree at a height of about 50 feet, the machine suddenly began to turn up on one wing, and took a course toward the tree. The operator, not relishing the idea of landing in a thorn tree, attempted to reach the ground. The left wing, however, struck the tree at a height of 10 or 12 feet from the ground and carried away several branches; but the flight, which had already covered a distance of six miles, was continued to the starting point.

'The causes of these troubles--too technical for explanation here--were not entirely overcome till the end of September, 1905. The flights then rapidly increased in length, till experiments were discontinued after October 5 on account of the number of people attracted to the field. Although made on a ground open on every side, and bordered on two sides by much-travelled thoroughfares, with electric cars passing every hour, and seen by all the people living in the neighbourhood for miles around, and by several hundred others, yet these flights have been made by some newspapers the subject of a great "mystery."'

Viewing their work from the financial side, the two brothers incurred but little expense in the earlier gliding experiments, and, indeed, viewed these only as recreation, limiting their expenditure to that which two men might spend on any hobby. When they had once achieved successful power-driven flight, they saw the possibilities of their work, and abandoned such other business as had engaged their energies, sinking all their capital in the development of a practical flying machine. Having, in 1905, improved their designs to such an extent that they could consider their machine a practical aeroplane, they devoted the years 1906 and 1907 to business negotiations and to the construction of new machines, resuming flying experiments in May of 1908 in order to test the ability of their machine to meet the requirements of a contract they had made with the United States Government, which required an aeroplane capable of carrying two men, together with sufficient fuel supplies for a flight of 125 miles at 40 miles per hour. Practically similar to the machine used in the experiments of 1905, the contract aeroplane was fitted with a larger motor, and provision was made for seating a passenger and also for allowing of the operator assuming a sitting position, instead of lying prone.

Before leaving the work of the brothers to consider contemporary events, it may be noted that they claimed--with justice--that they were first to construct wings adjustable to different angles of incidence on the right and left side in order to control the balance of an aeroplane; the first to attain lateral balance by adjusting wing-tips to respectively different angles of incidence on the right and left sides, and the first to use a vertical vane in combination with wing-tips, adjustable to respectively different angles of incidence, in balancing and steering an aeroplane. They were first, too, to use a movable vertical tail, in combination with wings adjustable to different angles of incidence, in controlling the balance and direction of an aeroplane[5].

A certain Henry M. Weaver, who went to see the work of the brothers, writing in a letter which was subsequently read before the Aero Club de France records that he had a talk in 1905 with the farmer who rented the field in which the Wrights made their flights.' On October 5th (1905) he was cutting corn in the next field east, which is higher ground. When he noticed the aeroplane had started on its flight he remarked to his helper: "Well, the boys are at it again," and kept on cutting corn, at the same time keeping an eye on the great white form rushing about its course. "I just kept on shocking corn," he continued, "until I got down to the fence, and the durned thing was still going round. I thought it would never stop."'

He was right. The brothers started it, and it will never stop.

Mr Weaver also notes briefly the construction of the 1905 Wright flier. 'The frame was made of larch wood-from tip to tip of the wings the dimension was 40 feet. The gasoline motor--a special construction made by them--much the same, though, as the motor on the Pope-Toledo automobile--was of from 12 to 15 horsepower. The motor weighed 240 lbs. The frame was covered with ordinary muslin

5 Aeronautical Journal, No. 79.

of good quality. No attempt was made to lighten the machine; they simply built it strong enough to stand the shocks. The structure stood on skids or runners, like a sleigh. These held the frame high enough from the ground in alighting to protect the blades of the propeller. Complete with motor, the machine weighed 925 lbs.

XII. THE FIRST YEARS OF CONQUEST

It is no derogation of the work accomplished by the Wright Brothers to say that they won the honour of the first power-propelled flights in a heavier-than-air machine only by a short period. In Europe, and especially in France, independent experiment was being conducted by Ferber, by Santos-Dumont, and others, while in England Cody was not far behind the other giants of those days. The history of the early years of controlled power flights is a tangle of half-records; there were no chroniclers, only workers, and much of what was done goes unrecorded perforce, since it was not set down at the time.

Before passing to survey of those early years, let it be set down that in 1907, when the Wright Brothers had proved the practicability of their machines, negotiations were entered into between the brothers and the British War office. On April 12th 1907, the apostle of military stagnation, Haldane, then War Minister, put an end to the negotiations by declaring that 'the War office is not disposed to enter into relations at present with any manufacturer of aeroplanes' The state of the British air service in 1914 at the outbreak of hostilities, is eloquent regarding the pursuance of the policy which Haldane initiated.

'If I talked a lot,' said Wilbur Wright once, 'I should be like the parrot, which is the bird that speaks most and flies least.' That attitude is emblematic of the majority of the early fliers, and because of it the record of their achievements is incomplete to-day. Ferber, for instance, has left little from which to state what he did, and that little is scattered through various periodicals, scrappily enough. A French army officer, Captain Ferber was experimenting with monoplane and biplane gliders at the beginning of the century-his work was contemporary with that of the Wrights. He corresponded both with Chanute and with the Wrights, and in the end he was commissioned by the French Ministry of War to undertake the journey to America in order to negotiate with the Wright Brothers concerning French rights in the patents they had acquired, and to study their work at first hand.

Ferber's experiments in gliding began in 1899 at the Military School at Fountainebleau, with a canvas glider of some 80 square feet supporting surface, and weighing 65 lbs. Two years later he constructed a larger and more satisfactory machine, with which he made numerous excellent glides. Later, he constructed an apparatus which suspended a plane from a long arm which swung on a tower, in order that experiments might be carried out without risk to the experimenter, and it was not until 1905 that he attempted power-driven free flight. He took up the Voisin design of biplane for his power-driven flights, and virtually devoted all his energies to the study of aeronautics. His book, Aviation, its Dawn and Development, is a work of scientific value--unlike many of his contemporaries, Ferber brought to the study of the problems of flight a trained mind, and he was concerned equally with the theoretical problems of aeronautics and the practical aspects of the subject.

After Bleriot's successful cross-Channel flight, it was proposed to offer a prize of L1,000 for the feat which C. S. Rolls subsequently accomplished (starting from the English side of the Channel), a flight from Boulogne to Dover and back; in place of this, however, an aviation week at Boulogne was organised, but, although numerous aviators were invited to compete, the condition of the flying grounds was such that no competitions took place. Ferber was virtually the only one to do any flying at Boulogne, and at the outset he had his first accident; after what was for those days a good flight, he made a series of circles with his machine, when it suddenly struck the ground, being partially wrecked. Repairs were carried out, and Ferber resumed his exhibition flights, carrying on up to Wednesday, September 22nd, 1909. On that day he remained in the air for half an hour, and, as he was about to land, the machine struck a mound of earth and overturned, pinning Ferber under the weight of the motor. After being extricated, Ferber seemed to show little concern at the accident, but in a few minutes he complained of great pain, when he was conveyed to the ambulance shed on the ground.

'I was foolish,' he told those who were with him there. 'I was flying too low. It was my own fault and it will be a severe lesson to me. I wanted to turn round, and was only five metres from the ground.' A little after this, he got up from the couch on which he had been placed, and almost immediately collapsed, dying five minutes later.

Ferber's chief contemporaries in France were Santos-Dumont, of airship fame, Henri and Maurice Farman, Hubert Latham, Ernest Archdeacon, and Delagrange. These are names that come at once to mind, as does that of Bleriot, who accomplished the second great feat of power-driven flight, but as a matter of fact the years 1903-10 are filled with a little host of investigators and experimenters, many of whom, although their names do not survive to any extent, are but a very little way behind those mentioned here in enthusiasm and devotion. Archdeacon and Gabriel Voisin, the former of whom took to heart the success achieved by the Wright Brothers, co-operated in experiments in gliding. Archdeacon constructed a glider in box-kite fashion, and Voisin experimented with it on the Seine, the glider being towed by a motorboat to attain the necessary speed. It was Archdeacon who offered a cup for the first straight flight of 200 metres, which was won by Santos-Dumont, and he also combined with Henri Deutsch de la Meurthe in giving the prize for the first circular flight of a mile, which was won by Henry Farman on January 13th, 1908.

A history of the development of aviation in France in these, the strenuous years, would fill volumes in itself. Bleriot was carrying out experiments with a biplane glider on the Seine, and Robert Esnault-Pelterie was working on the lines of the Wright Brothers, bringing American practice to France. In America others besides the Wrights had wakened to the possibilities of heavier-than-air flight; Glenn Curtiss, in company with Dr Alexander Graham Bell, with J. A. D. McCurdy, and with F. W. Baldwin, a Canadian engineer, formed the Aerial Experiment Company, which built a number of aeroplanes, most famous of which were the 'June Bug,' the 'Red Wing,' and the 'White Wing.' In 1908 the 'June Bug 'won a cup presented by the Scientific American--it was the first prize offered in America in connection with aeroplane flight.

Among the little group of French experimenters in these first years of practical flight, Santos-Dumont takes high rank. He built his 'No. 14 bis' aeroplane in biplane form, with two superposed main plane surfaces, and fitted it with an eight-cylinder Antoinette motor driving a two-bladed aluminium propeller, of which the blades were 6 feet only from tip to tip. The total lift surface of 860 square feet was given with a wing-span of a little under 40 feet, and the weight of the complete machine was 353 lbs., of which the engine weighed 158 lbs. In July of 1906 Santos-Dumont flew a distance of a few yards in this machine, but damaged it in striking the ground; on October 23rd of the same year he made a flight of nearly 200 feet--which might have been longer, but that he feared a crowd in front of the aeroplane and cut off his ignition. This may be regarded as the first effective flight in Europe, and by it Santos-Dumont takes his place as one of the chief--if not the chief--of the pioneers of the first years of practical flight, so far as Europe is concerned.

Meanwhile, the Voisin Brothers, who in 1904 made cellular kites for Archdeacon to test by towing on the Seine from a motor launch, obtained data for the construction of the aeroplane which Delagrange and Henry Farman were to use later. The Voisin was a biplane, constructed with due regard to the designs of Langley, Lilienthal, and other earlier experimenters--both the Voisins and M. Colliex, their engineer, studied Lilienthal pretty exhaustively in getting out their design, though their own researches were very thorough as well. The weight of this Voisin biplane was about 1,450 lbs., and its maximum speed was some 38 to 40 miles per hour, the total supporting surface being about 535 square feet. It differed from the Wright design in the possession of a tail-piece, a characteristic which marked all the French school of early design as in opposition to the American. The Wright machine got its longitudinal stability by means of the main planes and the elevating planes, while the Voisin type added a third factor of stability in its sailplanes. Further, the Voisins fitted their biplane with a wheeled undercarriage, while the Wright machine, being fitted only with runners, demanded a launching rail for starting. Whether a machine should be tailless or tailed was for some long time matter for acute controversy, which in the end was settled by the fitting of a tail to the Wright machines-France won the dispute by the concession.

Henry Farman, who began his flying career with a Voisin machine, evolved from it the aeroplane which bore his name, following the main lines of the Voisin type fairly closely, but making alterations in the controls, and in the design of the undercarriage, which was somewhat elaborated, even to the inclusion of shock absorbers. The seven-cylinder 50 horse-power Gnome rotary engine was fitted to the Farman machine--the Voisins had fitted an eight-cylinder Antoinette, giving 50 horse-power at 1,100 revolutions per minute, with direct drive to the propeller. Farman reduced the weight of the machine from the 1,450 lbs. of the Voisins to some 1,010 lbs. or thereabouts, and the supporting area to 450 square feet. This machine won its chief fame with Paulhan as pilot in the famous London to Manchester flight--it is to be remarked, too, that Farman himself was the first man in Europe to accomplish a flight of a mile.

Other notable designs of these early days were the 'R.E.P.', Esnault Pelterie's machine, and the Curtiss-Herring biplane. Of these Esnault Pelterie's was a monoplane, designed in that form since Esnault Pelterie had found by experiment that the wire used in bracing offers far more resistance to the air than its dimensions would seem to warrant. He built the wings of sufficient strength to stand the strain of flight without bracing wires, and dependent only for their support on the points of attachment to the body of the machine; for the rest, it carried its propeller in front of the planes, and both horizontal and vertical rudders at the stern--a distinct departure from the Wright and similar types. One wheel only was fixed under the body where the undercarriage exists on a normal design, but light wheels were fixed, one at the extremity of each wing, and there was also a wheel under the tail portion of the machine. A single lever actuated all the controls for steering. With a supporting surface of 150 square feet the machine weighed 946 lbs., about 6.4 lbs. per square foot of lifting surface.

The Curtiss biplane, as flown by Glenn Curtiss at the Rheims meeting, was built with a bamboo framework, stayed by means of very fine steel-stranded cables. A--then--novel feature of the machine was the moving of the ailerons by the pilot leaning to one side or the other in his seat, a light, tubular arm-rest being pressed by his body when he leaned to one side or the other, and thus operating the movement of the ailerons employed for tilting the plane when turning. A steering-wheel fitted immediately in front of the pilot's seat served to operate a rear steering-rudder when the wheel was turned in either direction, while pulling back the wheel altered the inclination of the front elevating planes, and so gave lifting or depressing control of the plane.

This machine ran on three wheels before leaving the ground, a central undercarriage wheel being fitted in front, with two more in line with a right angle line drawn through the centre of the engine crank at the rear end of the crank-case. The engine was a 35 horsepower Vee design, water cooled, with overhead inlet and exhaust valves, and Bosch high-tension magneto ignition. The total weight of the plane in flying order was about 700 lbs.

As great a figure in the early days as either Ferber or Santos-Dumont was Louis Bleriot, who, as early as 1900 built a flapping-wing model, this before ever he came to experimenting with the Voisin biplane type of glider on the Seine. Up to 1906 he had built four biplanes of his own design, and in March of 1907 he built his first monoplane, to wreck it only a few days after completion in an accident from which he had a fortunate escape. His next machine was a double monoplane, designed after Langley's precept, to a certain extent, and this was totally wrecked in September of 1907. His seventh machine, a monoplane, was built within a month of this accident, and with this he had a number of mishaps, also achieving some good flights, including one in which he made a turn. It was wrecked in December of 1907, whereupon he built another monoplane on which, on July 6th, 1908, Bleriot made a flight lasting eight and a half minutes. In October of that year he flew the machine from Toury to Artenay and returned on it--this was just a day after Farman's first cross-country flight--but, trying to repeat the success five days later, Bleriot collided with a tree in a fog and wrecked the machine past repair. Thereupon he set about building his eleventh machine, with which he was to achieve the first flight across the English channel.

Henry Farman, to whom reference has already been made, was engaged with his two brothers, Maurice and Richard, in the motor-car business, and turned to active interest in flying in 1907, when the Voisin firm built his first biplane on the box-kite principle. In July of 1908 he won a prize of L400 for a flight of thirteen miles, previously having completed the first kilometre flown in Europe with a passenger, the said passenger being Ernest Archdeaon. In September of 1908 Farman put up a speed record of forty miles an hour in a flight lasting forty minutes.

Santos-Dumont produced the famous 'Demoiselle' monoplane early in 1909, a tiny machine in which the pilot had his seat in a sort of miniature cage under the main plane. It was a very fast, light little machine but was difficult to fly, and owing to its small wingspread was unable to glide at a reasonably safe angle. There has probably never been a cheaper flying machine to build than the 'Demoiselle,' which could be so upset as to seem completely wrecked, and then repaired ready for further flight by a couple of hours' work. Santos-Dumont retained no patent in the design, but gave it out freely to any one who chose to build 'Demoiselles'; the vogue of the pattern was brief, owing to the difficulty of piloting the machine.

These were the years of records, broken almost as soon as made. There was Farman's mile, there was the flight of the Comte de Lambert over the Eiffel Tower, Latham's flight at Blackpool in a high wind, the Rheims records, and then Henry Farman's flight of four hours later in 1909, Orville Wright's height record of 1,640 feet, and Delagrange's speed record of 49.9 miles per hour. The coming to fame of the Gnome rotary engine helped in the making of these records to a very great extent, for in this engine was a prime mover which gave the reliability that aeroplane builders and pilots had been searching for, but vainly. The Wrights and Glenn Curtiss, of course, had their own designs of engine, but the Gnome, in spite of its lack of economy in fuel and oil, and its high cost, soon came to be regarded as the best power plant for flight.

Delagrange, one of the very good pilots of the early days, provided a curious insight to the way in which flying was regarded, at the opening of the Juvisy aero aerodrome in May of 1909. A huge crowd had gathered for the first day's flying, and nine machines were announced to appear, but only three were brought out. Delagrange made what was considered an indifferent little flight, and another pilot, one De Bischoff, attempted to rise, but could not get his machine off the ground. Thereupon the crowd of 30,000 people lost their tempers, broke down the barriers surrounding the flying course, and hissed the officials, who were quite unable to maintain order. Delagrange, however, saved the situation by making a circuit of the course at a height of thirty feet from the ground, which won him rounds of cheering and restored the crowd to good humour. Possibly the smash achieved by Rougier, the famous racing motorist, who crashed his Voisin biplane after Delagrange had made his circuit, completed the enjoyment of the spectators. Delagrange, flying at Argentan in June of 1909, made a flight of four kilometres at a height of sixty feet; for those days this was a noteworthy performance. Contemporary with this was Hubert Latham's flight of an hour and seven minutes on an Antoinette monoplane; this won the adjective 'magnificent' from contemporary recorders of aviation.

Viewing the work of the little group of French experimenters, it is, at this length of time from their exploits, difficult to see why they carried the art as far as they did. There was in it little of satisfaction, a certain measure of fame, and practically no profit--the giants of those days got very little for their pains. Delagrange's experience at the opening of the Juvisy ground was symptomatic of the way in which flight was regarded by the great mass of people--it was a sport, and nothing more, but a sport without the dividends attaching to professional football or horse-racing. For a brief period, after the Rheims meeting, there was a golden harvest to be reaped by the best of the pilots. Henry Farman asked L2,000 for a week's exhibition flying in England, and Paulhan asked half that sum, but a rapid increase in the number of capable pilots, together with the fact that most flying meetings were financial failures, owing to great expense in organisation and the doubtful factor of the weather, killed this goose before many golden eggs had been gathered in by the star aviators. Besides, as height and distance records were broken one after another, it became less and less necessary to pay for entrance to an aerodrome in order to see a flight--the thing grew too big for a mere sports ground.

Long before Rheims and the meeting there, aviation had grown too big for the chronicling of every individual effort. In that period of the first days of conquest of the air, so much was done by so many whose names are now half-forgotten that it is possible only to pick out the great figures and make brief reference to their achievements and the machines with which they accomplished so much, pausing to note such epoch-making events as the London-Manchester flight, Bleriot's Channel crossing, and the Rheims meeting itself, and then passing on beyond the days of individual records to the time when the machine began to dominate the man. This latter because, in the early days, it was heroism to trust life to the planes that were turned out--the 'Demoiselle' and the Antoinette machine that Latham used in his attempt to fly the Channel are good examples of the flimsiness of early types--while in the later period, that of the war and subsequently, the heroism turned itself in a different--and nobler-direction. Design became standardised, though not perfected. The domination of the machine may best be expressed by contrasting the way in which machines came to be regarded as compared with the men who flew them: up to 1909, flying enthusiasts talked of Farman, of Bleriot, of Paulhan, Curtiss, and of other men; later, they began to talk of the Voisin, the Deperdussin, and even to the Fokker, the Avro, and the Bristol type. With the standardising of the machine, the days of the giants came to an end.

XIII. FIRST FLIERS IN ENGLAND

Certain experiments made in England by Mr Phillips seem to have come near robbing the Wright Brothers of the honour of the first flight; notes made by Colonel J. D. Fullerton on the Phillips flying machine show that in 1893 the first machine was built with a length of 25 feet, breadth of 22 feet, and height of 11 feet, the total weight, including a 72 lb. load, being 420 lbs. The machine was fitted with some fifty wood slats, in place of the single supporting surface of the monoplane or two superposed surfaces of the biplane, these slats being fixed in a steel frame so that the whole machine rather resembled a Venetian blind. A steam engine giving about 9 horse-power provided the motive power for the six-foot diameter propeller which drove the machine. As it was not possible to put a passenger in control as pilot, the machine was attached to a central post by wire guys and run round a circle 100 feet in diameter, the track consisting of wooden planking 4 feet wide. Pressure of air under the slats caused the machine to rise some two or three feet above the track when sufficient velocity had been attained, and the best trials were made on June 19th 1893, when at a speed of 40 miles an hour, with a total load of 385 lbs., all the wheels were off the ground for a distance of 2,000 feet.

In 1904 a full-sized machine was constructed by Mr Phillips, with a total weight, including that of the pilot, of 600 lbs. The machine was designed to lift when it had attained a velocity of 50 feet per second, the motor fitted giving 22 horse-power. On trial, however, the longitudinal equilibrium was found to be defective, and a further design was got out, the third machine being completed in 1907. In this the wood slats were held in four parallel container frames, the weight of the machine, excluding the pilot, being 500 lbs. A motor similar to that used in the 1904 machine was fitted, and the machine was designed to lift at a velocity of about 30 miles an hour, a seven-foot propeller doing the driving. Mr Phillips tried out this machine in a field about 400 yards across. 'The machine was started close to the hedge, and rose from the ground when about 200 yards had been covered. When the machine touched the ground again, about which there could be no doubt, owing to the terrific jolting, it did not run many yards. When it came to rest I was about ten yards from the boundary. Of course, I stopped the engine before I commenced to descend.'6

S. F. Cody, an American by birth, aroused the attention not only of the British public, but of the War office and Admiralty as well, as early as 1905 with his man-lifting kites. In that year a height of 1,600 feet was reached by one of these box-kites, carrying a man, and later in the same year one Sapper Moreton, of the Balloon Section of the Royal Engineers (the parent of the Royal Flying Corps) remained for an hour at an altitude of 2,600 feet. Following on the success of these kites, Cody constructed an aeroplane which he designated a 'power kite,' which was in reality a biplane that made the first flight in Great Britain. Speaking before the Aeronautical Society in 1908, Cody said that 'I have accomplished one thing that I hoped for very much, that is, to be the first man to fly in Great Britain.... I made a machine that left the ground the first time out; not high, possibly five or six inches only. I might have gone higher if I wished. I made some five flights in all, and the last flight came to grief.... On the morning of the accident I went out after adjusting my propellers at 8 feet pitch running at 600 (revolutions per minute). I think that I flew at about twenty-eight miles per hour. I had 50 horsepower motor power in the engine. A bunch of trees, a flat common above these trees, and from this flat there is a slope goes down... to another clump of trees. Now, these clumps of trees are a quarter of a mile apart or thereabouts.... I was accused of doing nothing but jumping with my machine, so I got a bit agitated and went to fly.

I went out this morning with an easterly wind, and left the ground at the bottom of the hill and struck the ground at the top, a distance of 74 yards. That proved beyond a doubt that the machine would fly--it flew uphill. That was the most talented flight the machine did, in my opinion. Now, I turned round at the top and started the machine and left the ground--remember, a ten mile wind was blowing at the time. Then, 60 yards from where the men let go, the machine went off in this direction (demonstrating)--I make a line now where I hoped to land--to cut these trees off at that side and land right off in here. I got here somewhat excited, and started down and saw these trees right in front of me. I did not want to smash my head rudder to pieces, so I raised it again and went up. I got one wing direct over that clump of trees, the right wing over the trees, the left wing free; the wind, blowing with me, had to lift over these trees. So I consequently got a false lift on

---
6 Aeronautical Journal, July, 1908.

the right side and no lift on the left side. Being only about 8 feet from the tree tops, that turned my machine up like that (demonstrating). This end struck the ground shortly after I had passed the trees. I pulled the steering handle over as far as I could. Then I faced another bunch of trees right in front of me. Trying to avoid this second bunch of trees I turned the rudder, and turned it rather sharp. That side of the machine struck, and it crumpled up like so much tissue paper, and the machine spun round and struck the ground that way on, and the framework was considerably wrecked. Now, I want to advise all aviators not to try to fly with the wind and to cross over any big clump of earth or any obstacle of any description unless they go square over the top of it, because the lift is enormous crossing over anything like that, and in coming the other way against the wind it would be the same thing when you arrive at the windward side of the obstacle. That is a point I did not think of, and had I thought of it I would have been more cautious.'

This Cody machine was a biplane with about 40 foot span, the wings being about 7 feet in depth with about 8 feet between upper and lower wing surfaces. 'Attached to the extremities of the lower planes are two small horizontal planes or rudders, while a third small vertical plane is fixed over the centre of the upper plane.' The tail-piece and principal rudder were fitted behind the main body of the machine, and a horizontal rudder plane was rigged out in front, on two supporting arms extending from the centre of the machine. The small end-planes and the vertical plane were used in conjunction with the main rudder when turning to right or left, the inner plane being depressed on the turn, and the outer one correspondingly raised, while the vertical plane, working in conjunction, assisted in preserving stability. Two two-bladed propellers were driven by an eight-cylinder 50 horse-power Antoinette motor. With this machine Cody made his first flights over Laffan's plain, being then definitely attached to the Balloon Section of the Royal Engineers as military aviation specialist.

There were many months of experiment and trial, after the accident which Cody detailed in the statement given above, and then, on May 14th, 1909, Cody took the air and made a flight of 1,200 yards with entire success. Meanwhile A. V. Roe was experimenting at Lea Marshes with a triplane of rather curious design the pilot having his seat between two sets of three superposed planes, of which the front planes could be tilted and twisted while the machine was in motion. He comes but a little way after Cody in the chronology of early British experimenters, but Cody, a born inventor, must be regarded as the pioneer of the present century so far as Britain is concerned. He was neither engineer nor trained mathematician, but he was a good rule-of-thumb mechanic and a man of pluck and perseverance; he never strove to fly on an imperfect machine, but made alteration after alteration in order to find out what was improvement and what was not, in consequence of which it was said of him that he was 'always satisfied with his alterations.'

By July of 1909 he had fitted an 80 horse-power motor to his biplane, and with this he made a flight of over four miles over Laffan's Plain on July 21st. By August he was carrying passengers, the first being Colonel Capper of the R.E. Balloon Section, who flew with Cody for over two miles, and on September 8th, 1909, he made a world's record cross-country flight of over forty miles in sixty-six minutes, taking a course from Laffan's Plain over Farnborough, Rushmoor, and Fleet, and back to Laffan's Plain. He was one of the competitors in the 1909 Doncaster Aviation Meeting, and in 1910 he competed at Wolverhampton, Bournemouth, and Lanark. It was on June 7th, 1910, that he qualified for his brevet, No. 9, on the Cody biplane.

He built a machine which embodied all the improvements for which he had gained experience, in 1911, a biplane with a length of 35 feet and span of 43 feet, known as the 'Cody cathedral' on account of its rather cumbrous appearance. With this, in 1911, he won the two Michelin trophies presented in England, completed the Daily Mail circuit of Britain, won the Michelin cross-country prize in 1912 and altogether, by the end of 1912, had covered more than 7,000 miles with the machine. It was fitted with a 120 horse-power Austro-Daimler engine, and was characterised by an exceptionally wide range of speed--the great wingspread gave a slow landing speed.

A few of his records may be given: in 1910, flying at Laffan's Plain in his biplane, fitted with a 50-60 horsepower Green engine, on December 31st, he broke the records for distance and time by flying 185 miles, 787 yards, in 4 hours 37 minutes. On October 31st, 1911, he beat this record by flying for 5 hours 15 minutes, in which period he covered 261 miles 810 yards with a 60 horse-power Green engine fitted to his biplane. In 1912, competing in the British War office tests of military aeroplanes, he won the L5,000 offered by the War Office. This was in competition with no less than twenty-five other machines, among which were the since-famous

Deperdussin, Bristol, Flanders, and Avro types, as well as the Maurice Farman and Bleriot makes of machine. Cody's remarkable speed range was demonstrated in these trials, the speeds of his machine varying between 72.4 and 48.5 miles per hour. The machine was the only one delivered for the trials by air, and during the three hours' test imposed on all competitors a maximum height of 5,000 feet was reached, the first thousand feet being achieved in three and a half minutes.

During the summer of 1913 Cody put his energies into the production of a large hydro-biplane, with which he intended to win the L5,000 prize offered by the Daily Mail to the first aviator to fly round Britain on a waterplane. This machine was fitted with landing gear for its tests, and, while flying it over Laffan's Plain on August 7th, 1913, with Mr W. H. B. Evans as passenger, Cody met with the accident that cost both him and his passenger their lives. Aviation lost a great figure by his death, for his plodding, experimenting, and dogged courage not only won him the fame that came to a few of the pilots of those days, but also advanced the cause of flying very considerably and contributed not a little to the sum of knowledge in regard to design and construction.

Another figure of the early days was A. V. Roe, who came from marine engineering to the motor industry and aviation in 1905. In 1906 he went out to Colorado, getting out drawings for the Davidson helicopter, and in 1907 having returned to England, he obtained highest award out of 200 entries in a model aeroplane flying competition. From the design of this model he built a full-sized machine, and made a first flight on it, fitted with a 24 horse-power Antoinette engine, in June of 1908 Later, he fitted a 9 horsepower motor-cycle engine to a triplane of his own design, and with this made a number of short flights; he got his flying brevet on a triplane with a motor of 35 horse-power, which, together with a second triplane, was entered for the Blackpool aviation meeting of 1910 but was burnt in transport to the meeting. He was responsible for the building of the first seaplane to rise from English waters, and may be counted the pioneer of the tractor type of biplane. In 1913 he built a two-seater tractor biplane with 80 horse-power engine, a machine which for some considerable time ranked as a leader of design. Together with E. V. Roe and H. V. Roe, 'A. V.' controlled the Avro works, which produced some of the most famous training machines of the war period in a modification of the original 80 horse-power tractor. The first of the series of Avro tractors to be adopted by the military authorities was the 1912 biplane, a two-seater fitted with 50 horsepower engine. It was the first tractor biplane with a closed fuselage to be used for military work, and became standard for the type. The Avro seaplane, of I 100 horse-power (a fourteen-cylinder Gnome engine was used) was taken up by the British Admiralty in 1913. It had a length of 34 feet and a wing-span of 50 feet, and was of the twin-float type.

Geoffrey de Havilland, though of later rank, counts high among designers of British machines. He qualified for his brevet as late as February, 1911, on a biplane of his own construction, and became responsible for the design of the BE2, the first successful British Government biplane. On this he made a British height record of 10,500 feet over Salisbury Plain, in August of 1912, when he took up Major Sykes as passenger. In the war period he was one of the principal designers of fighting and reconnaissance machines.

F. Handley Page, who started in business as an aeroplane builder in 1908, having works at Barking, was one of the principal exponents of the inherently stable machine, to which he devoted practically all his experimental work up to the outbreak of war. The experiments were made with various machines, both of monoplane and biplane type, and of these one of the best was a two-seater monoplane built in 1911, while a second was a larger machine, a biplane, built in 1913 and fitted with a 110 horse-power Anzani engine. The war period brought out the giant biplane with which the name of Handley Page is most associated, the twin-engined night-bomber being a familiar feature of the later days of the war; the four-engined bomber had hardly had a chance of proving itself under service conditions when the war came to an end.

Another notable figure of the early period was 'Tommy' Sopwith, who took his flying brevet at Brooklands in November of 1910, and within four days made the British duration record of 108 miles in 3 hours 12 minutes. On December 18th, 1910, he won the Baron de Forrest prize of L4,000 for the longest flight from England to the Continent, flying from Eastchurch to Tirlemont, Belgium, in three hours, a distance of 161 miles. After two years of touring in America, he returned to England and established a flying school. In 1912 he won the first aerial Derby, and in 1913 a machine of his design, a tractor biplane, raised the British height record to 13,000 feet (June 16th, at Brooklands). First as aviator, and then as designer, Sopwith has done much useful work in aviation.

These are but a few, out of a host who contributed to the development of flying in this country, for, although France may be said to have set the pace as regards development, Britain was not far behind. French experimenters received far more Government aid than did the early British aviators and designers--in the early days the two were practically synonymous, and there are many stories of the very early days at Brooklands, where, when funds ran low, the ardent spirits patched their trousers with aeroplane fabric and went on with their work with Bohemian cheeriness. Cody, altering and experimenting on Laffan's Plain, is the greatest figure of them all, but others rank, too, as giants of the early days, before the war brought full recognition of the aeroplane's potentialities.

One of the first men actually to fly in England, Mr J. C. T. Moore-Brabazon, was a famous figure in the days of exhibition flying, and won his reputation mainly through being first to fly a circular mile on a machine designed and built in Great Britain and piloted by a British subject. Moore-Brabazon's earliest flights were made in France on a Voisin biplane in 1908, and he brought this machine over to England, to the Aero Club grounds at Shellness, but soon decided that he would pilot a British machine instead. An order was placed for a Short machine, and this, fitted with a 50-60 horse-power Green engine, was used for the circular mile, which won a prize of L1,000 offered by the Daily Mail, the feat being accomplished on October 30th, 1909. Five days later, Moore-Brabazon achieved the longest flight up to that time accomplished on a British-built machine, covering three and a half miles. In connection with early flying in England, it is claimed that A. V. Roe, flying 'Avro B,',' on June 8th, 1908, was actually the first man to leave the ground, this being at Brooklands, but in point of fact Cody antedated him.

No record of early British fliers could be made without the name of C. S. Rolls, a son of Lord Llangattock, on June 2nd, 1910, he flew across the English Channel to France, until he was duly observed over French territory, when he returned to England without alighting. The trip was made on a Wright biplane, and was the third Channel crossing by air, Bleriot having made the first, and Jacques de Lesseps the second. Rolls was first to make the return journey in one trip. He was eventually killed through the breaking of the tail-plane of his machine in descending at a flying meeting at Bournemouth. The machine was a Wright biplane, but the design of the tail-plane--which, by the way, was an addition to the machine, and was not even sanctioned by the Wrights--appears to have been carelessly executed, and the plane itself was faulty in construction. The breakage caused the machine to overturn, killing Rolls, who was piloting it.

XIV. RHEIMS, AND AFTER

The foregoing brief--and necessarily incomplete--survey of the early British group of fliers has taken us far beyond some of the great events of the early days of successful flight, and it is necessary to go back to certain landmarks in the history of aviation, first of which is the great meeting at Rheims in 1909. Wilbur Wright had come to Europe, and, flying at Le Mans and Pau--it was on August 8th, 1908, that Wilbur Wright made the first of his ascents in Europe--had stimulated public interest in flying in France to a very great degree. Meanwhile, Orville Wright, flying at Fort Meyer, U.S.A., with Lieutenant Selfridge as a passenger, sustained an accident which very nearly cost him his life through the transmission gear of the motor breaking. Selfridge was killed and Orville Wright was severely injured--it was the first fatal accident with a Wright machine.

Orville Wright made a flight of over an hour on September 9th, 1908, and on December 31st of that year Wilbur flew for 2 hours 19 minutes. Thus, when the Rheims meeting was organised--more notable because it was the first of its kind, there were already records waiting to be broken. The great week opened on August 22nd, there being thirty entrants, including all the most famous men among the early fliers in France. Bleriot, fresh from his Channel conquest, was there, together with Henry Farman, Paulhan, Curtiss, Latham, and the Comte de Lambert, first pupil of the Wright machine in Europe to achieve a reputation as an aviator.

'To say that this week marks an epoch in the history of the world is to state a platitude. Nevertheless, it is worth stating, and for us who are lucky enough to be at Rheims during this week there is a solid satisfaction in the idea that we are present at the making of history. In perhaps only a few years to come the competitions of this week may look pathetically small and the distances and speeds may appear paltry. Nevertheless, they are the first of their kind, and that is sufficient.'

So wrote a newspaper correspondent who was present at the famous meeting, and his words may stand, being more than mere journalism; for the great flying week which opened on August 22nd, 1909, ranks as one of the great landmarks in the history of heavier-than-air flight. The day before the opening of the meeting a downpour of rain spoilt the flying ground; Sunday opened with a fairly high wind, and in a lull M. Guffroy turned out on a crimson R.E.P. monoplane, but the wheels of his undercarriage stuck in the mud and prevented him from rising in the quarter of an hour allowed to competitors to get off the ground. Bleriot, following, succeeded in covering one side of the triangular course, but then came down through grit in the carburettor. Latham, following him with thirteen as the number of his machine, experienced his usual bad luck and came to earth through engine trouble after a very short flight. Captain Ferber, who, owing to military regulations, always flew under the name of De Rue, came out next with his Voisin biplane, but failed to get off the ground; he was followed by Lefebvre on a Wright biplane, who achieved the success of the morning by rounding the course--a distance of six and a quarter miles--in nine minutes with a twenty mile an hour wind blowing. His flight finished the morning.

Wind and rain kept competitors out of the air until the evening, when Latham went up, to be followed almost immediately by the Comte de Lambert. Sommer, Cockburn (the only English competitor), Delagrange, Fournier, Lefebvre, Bleriot, Bunau-Varilla, Tissandier, Paulhan, and Ferber turned out after the first two, and the excitement of the spectators at seeing so many machines in the air at one time provoked wild cheering. The only accident of the day came when Bleriot damaged his propeller in colliding with a haycock.

The main results of the day were that the Comte de Lambert flew 30 kilometres in 29 minutes 2 seconds; Lefebvre made the ten-kilometre circle of the track in just a second under 9 minutes, while Tissandier did it in 9 1/4 minutes, and Paulhan reached a height of 230 feet. Small as these results seem to us now, and ridiculous as may seem enthusiasm at the sight of a few machines in the air at the same time, the Rheims Meeting remains a great event, since it proved definitely to the whole world that the conquest of the air had been achieved.

Throughout the week record after record was made and broken. Thus on the Monday, Lefebvre put up a record for rounding the course and Bleriot beat it, to be beaten in turn by Glenn Curtiss on his Curtiss-Herring biplane. On that day, too, Paulhan covered 34 3/4 miles in 1 hour 6 minutes. On the next day, Paulhan on his Voisin biplane took the air with Latham, and Fournier followed, only to smash up his machine by striking an eddy of wind which turned him over several times. On the Thursday, one of the chief events was Latham's 43 miles accomplished in 1 hour 2 minutes in the morning and his 96.5 miles in 2 hours 13 minutes in the afternoon, the latter flight only terminated by running out of petrol. On the Friday, the Colonel Renard French airship, which had flown over the ground under the pilotage of M. Kapfarer, paid Rheims a second visit; Latham manoeuvred round the airship on his Antoinette and finally left it far behind. Henry Farman won the Grand Prix de Champagne on this day, covering 112 miles in 3 hours, 4 minutes, 56 seconds, Latham being second with his 96.5 miles flight, and Paulhan third.

On the Saturday, Glenn Curtiss came to his own, winning the Gordon-Bennett Cup by covering 20 kilometres in 15 minutes 50.6 seconds. Bleriot made a good second with 15 minutes 56.2 seconds as his time, and Latham and Lefebvre were third and fourth. Farman carried off the passenger prize by carrying two passengers a distance of 6 miles in 10 minutes 39 seconds. On the last day Delagrange narrowly escaped serious accident through the bursting of his propeller while in the air, Curtiss made a new speed record by travelling at the rate of over 50 miles an hour, and Latham, rising to 500 feet, won the altitude prize.

These are the cold statistics of the meeting; at this length of time it is difficult to convey any idea of the enthusiasm of the crowds over the achievements of the various competitors, while the incidents of the week, comic and otherwise, are nearly forgotten now even by those present in this making of history. Latham's great flight on the Thursday was rendered a breathless episode by a downpour of rain when he had covered all but a kilometre of the record distance previously achieved by Paulhan, and there was wild enthusiasm when Latham flew on through the rain until he had put up a new record and his petrol had run out. Again, on the Friday afternoon, the Colonel Renard took the air together with a little French dirigible, Zodiac III; Latham was already in the air directly over Farman, who was also flying, and three crows which turned out as rivals to the human aviators received as much cheering for their appearance as had been accorded to the machines, which doubtless they could not understand. Frightened by the cheering, the crows tried to escape from the course, but as they came near the stands, the crowd rose to cheer again and the crows wheeled away to make a second charge towards safety, with the same result; the crowd rose and cheered at them a third and fourth time; between ten and fifteen thousand people stood on chairs and tables and waved hats and handkerchiefs at three ordinary, everyday crows. One thoughtful spectator, having thoroughly enjoyed the funny side of the incident, remarked that the ultimate mastery of the air lies with the machine that comes nearest to natural flight. This still remains for the future to settle.

Farman's world record, which won the Grand Prix de Champagne, was done with a Gnome Rotary Motor which had only been run on the test bench and was fitted to his machine four hours before he started on the great flight. His propeller had never been tested, having only been completed the night before. The closing laps of that flight, extending as they did into the growing of the dusk, made a breathlessly eerie experience for such of the spectators as stayed on to watch--and these were many. Night came on steadily and Farman covered lap after lap just as steadily, a buzzing, circling mechanism with something relentless in its isolated persistency.

The final day of the meeting provided a further record in the quarter million spectators who turned up to witness the close of the great week. Bleriot, turning out in the morning, made a landing in some such fashion as flooded the carburettor and caused it to catch fire. Bleriot himself was badly burned, since the petrol tank burst and, in the end, only the metal parts of the machine were left. Glenn Curtis tried to beat Bleriot's time for a lap of the course, but failed. In the evening, Farman and Latham went out and up in great circles, Farman cleaving his way upward in what at the time counted for a huge machine, on circles of about a mile diameter. His first round took him level with the top of the stands, and, in his second, he circled the captive balloon anchored in the middle of the grounds. After another circle, he came down on a long glide, when Latham's lean Antoinette monoplane went up in circles more graceful than those of Farman. 'Swiftly it rose and swept round close to the balloon, veered round to the hangars, and out over to the Rheims road. Back it came high over the stands, the people craning their necks as the shrill cry of the engine drew nearer and nearer behind the stands. Then of a sudden, the little form appeared away up in the deep twilight blue vault of the sky, heading straight as an arrow for the anchored balloon. Over it, and high, high above it went the Antoinette, seemingly higher by many feet than the Farman machine. Then, wheeling in a long sweep to the left, Latham steered his machine round past the stands, where the people, their nerve-tension released on seeing the machine descending from its perilous height of 500 feet, shouted their frenzied acclamations to the hero of the meeting.

'For certainly "Le Tham," as the French call him, was the popular hero. He always flew high, he always flew well, and his machine was a joy to the eye, either afar off or at close quarters. The public feeling for Bleriot is different. Bleriot, in the popular estimation, is the man who fights against odds, who meets the adverse fates calmly and with good courage, and to whom good luck comes once in a while as a reward for much labour and anguish, bodily and mental. Latham is the darling of the Gods, to whom Fate has only been unkind in the matter of the Channel flight, and only then because the honour belonged to Bleriot.

'Next to these two, the public loved most Lefebvre, the joyous, the gymnastic. Lefebvre was the comedian of the meeting. When things began to flag, the gay little Lefebvre would trot out to his starting rail, out at the back of the judge's enclosure opposite the stands, and after a little twisting of propellers his Wright machine would bounce off the end of its starting rail and proceed to do the most marvellous tricks for the benefit of the crowd, wheeling to right and left, darting up and down, now flying over a troop of the cavalry who kept the plain clear of people and sending their horses into hysterics, anon making straight for an unfortunate photographer who would throw himself and his precious camera flat on the ground to escape annihilation as Lefebvre swept over him 6 or 7 feet off the ground. Lefebvre was great fun, and when he had once found that his machine was not fast enough to compete for speed with the Bleriots, Antoinettes, and Curtiss, he kept to his metier of amusing people. The promoters of the meeting owe Lefebvre a debt of gratitude, for he provided just the necessary comic relief.'--(The Aero, September 7th, 1909.)

It may be noted, in connection with the fact that Cockburn was the only English competitor at the meeting, that the Rheims Meeting did more than anything which had preceded it to waken British interest in aviation. Previously, heavier-than-air flight in England had been regarded as a freak business by the great majority, and the very few pioneers who persevered toward winning England a share in the conquest of the air came in for as much derision as acclamation. Rheims altered this; it taught the world in general, and England in particular, that a serious rival to the dirigible balloon had come to being, and it awakened the thinking portion of the British public to the fact that the aeroplane had a future.

The success of this great meeting brought about a host of imitations of which only a few deserve bare mention since, unlike the first, they taught nothing and achieved little. There was the meeting at Boulogne late in September of 1909, of which the only noteworthy event was Ferber's death. There was a meeting at Brescia where Curtiss again took first prize for speed and Rougier put up a world's height record of 645 feet. The Blackpool meeting followed between 18th and 23rd of October, 1909, forming, with the exception of Doncaster, the first British Flying Meeting. Chief among the competitors were Henry Farman, who took the distance prize, Rougier, Paulhan, and Latham, who, by a flight in a high wind, convinced the British public that the theory that flying was only possible in a calm was a fallacy. A meeting at Doncaster was practically simultaneous with the Blackpool week; Delagrange, Le Blon, Sommer, and Cody were the principal figures in this event. It should be added that 130 miles was recorded as the total flown at Doncaster, while at Blackpool only 115 miles were flown. Then there were Juvisy, the first Parisian meeting, Wolverhampton, and the Comte de Lambert's flight round the Eiffel Tower at a height estimated at between 1,200 and 1,300 feet. This may be included in the record of these aerial theatricals, since it was nothing more.

Probably wakened to realisation of the possibilities of the aeroplane by the Rheims Meeting, Germany turned out its first plane late in 1909. It was known as the Grade monoplane, and was a blend of the Bleriot and Santos-Dumont machines, with a tail suggestive of the Antoinette type. The main frame took the form of a single steel tube, at the forward end of which was rigged a triangular arrangement carrying the pilot's seat and the landing wheels underneath, with the wing warping wires and stays above. The sweep of the wings was rather similar to the later Taube design, though the sweep back was not so pronounced, and the machine was driven by a four-cylinder, 20 horse-power, air-cooled engine which drove a two-bladed tractor propeller. In spite of Lilienthal's pioneer work years before, this was the first power-driven German plane which actually flew.

Eleven months after the Rheims meeting came what may be reckoned the only really notable aviation meeting on English soil, in the form of the Bournemouth week, July 10th to 16th, 1910. This gathering is noteworthy mainly in view of the amazing advance which it registered on the Rheims performances. Thus, in the matter of altitude, Morane reached 4,107 feet and Drexel came second with 2,490 feet. Audemars on a Demoiselle monoplane made a flight of 17 miles 1,480 yards in 27 minutes 17.2 seconds, a great flight for the little Demoiselle. Morane achieved a speed of 56.64 miles per hour, and Grahame White climbed to 1,000 feet altitude in 6 minutes 36.8 seconds. Machines carrying the Gnome engine as power unit took the great bulk of the prizes, and British-built engines were far behind.

The Bournemouth Meeting will always be remembered with regret for the tragedy of C. S. Rolls's death, which took place on the Tuesday, the second day of the meeting. The first competition of the day was that for the landing prize; Grahame White, Audemars, and Captain Dickson had landed with varying luck, and Rolls, following on a Wright machine with a tail-plane which ought never to have been fitted and was not part of the Wright design, came down wind after a left-hand turn and turned left again over the top of the stands in order to land up wind. He began to dive when just clear of the stands, and had dropped to a height of 40 feet when he came over the heads of the people against the barriers. Finding his descent too steep, he pulled back his elevator lever to bring the nose of the machine up, tipping down the front end of the tail to present an almost flat surface to the wind. Had all gone well, the nose of the machine would have been forced up, but the strain on the tail and its four light supports was too great; the tail collapsed, the wind pressed down the biplane elevator, and the machine dived vertically for the remaining 20 feet of the descent, hitting the ground vertically and crumpling up. Major Kennedy, first to reach the debris, found Rolls lying with his head doubled under him on the overturned upper main plane; the lower plane had been flung some few feet away with the engine and tanks under it. Rolls was instantaneously killed by concussion of the brain.

Antithesis to the tragedy was Audemars on his Demoiselle, which was named 'The Infuriated Grasshopper.' Concerning this, it was recorded at the time that 'Nothing so excruciatingly funny as the action of this machine has ever been seen at any aviation ground. The little two-cylinder engine pops away with a sound like the frantic drawing of ginger beer corks; the machine scutters along the ground with its tail well up; then down comes the tail suddenly and seems to slap the ground while the front jumps up, and all the spectators rock with laughter. The whole attitude and the jerky action of the machine suggest a grasshopper in a furious rage, and the impression is intensified when it comes down, as it did twice on Wednesday, in long grass, burying its head in the ground in its temper.'--(The Aero, July, 1910.)

The Lanark Meeting followed in August of the same year, and with the bare mention of this, the subject of flying meetings may he left alone, since they became mere matters of show until there came military competitions such as the Berlin Meeting at the end of August, 1910, and the British War office Trials on Salisbury Plain, when Cody won his greatest triumphs. The Berlin meeting proved that, from the time of the construction of the first successful German machine mentioned above, to the date of the meeting, a good number of German aviators had qualified for flight, but principally on Wright and Antoinette machines, though by that time the Aviatik and Dorner German makes had taken the air. The British War office Trials deserve separate and longer mention.

In 1910 in spite of official discouragement, Captain Dickson proved the value of the aeroplane for scouting purposes by observing movements of troops during the Military Manoeuvres on Salisbury Plain. Lieut. Lancelot Gibbs and Robert Loraine, the actor-aviator, also made flights over the manoeuvre area, locating troops and in a way anticipating the formation and work of the Royal Flying Corps by a usefulness which could not be officially recognised.

XV. THE CHANNEL CROSSING

It may be said that Louis Bleriot was responsible for the second great landmark in the history of successful flight. The day when the brothers Wright succeeded in accomplishing power-driven flight ranks as the first of these landmarks. Ader may or may not have left the ground, but the wreckage of his 'Avion' at the end of his experiment places his doubtful success in a different category from that of the brothers Wright and leaves them the first definite conquerors, just as Bleriot ranks as first definite conqueror of the English Channel by air.

In a way, Louis Bleriot ranks before Farman in point of time; his first flapping-wing model was built as early as 1900, and Voisin flew a biplane glider of his on the Seine in the very early experimental days. Bleriot's first four machines were biplanes, and his fifth, a monoplane, was wrecked almost immediately after its construction. Bleriot had studied Langley's work to a certain extent, and his sixth construction was a double monoplane based on the Langley principle. A month after he had wrecked this without damaging himself--for Bleriot had as many miraculous escapes as any of the other fliers-he brought out number seven, a fairly average monoplane. It was in December of 1907 after a series of flights that he wrecked this machine, and on its successor, in July of 1908, he made a flight of over 8 minutes. Sundry flights, more or less successful, including the first cross-country flight from Toury to Artenay, kept him busy up to the beginning of November, 1908, when the wreckage in a fog of the machine he was flying sent him to the building of 'number eleven,' the famous cross-channel aeroplane.

Number eleven was shown at the French Aero Show in the Grand Palais and was given its first trials on the 18th January, 1909. It was first fitted with a R.E.P. motor and had a lifting area of 120 square feet, which was later increased to 150 square feet. The framework was of oak and poplar spliced and reinforced with piano wire; the weight of the machine was 47 lbs. and the undercarriage weight a further 60 lbs., this consisting of rubber cord shock absorbers mounted on two wheels. The R.E.P. motor was found unsatisfactory, and a three-cylinder Anzani of 105 mm. bore and 120 mm. stroke replaced it. An accident seriously damaged the machine on June 2nd, but Bleriot repaired it and tested it at Issy, where between June 19th and June 23rd he accomplished flights of 8, 12, 15, 16, and 36 minutes. On July 4th he made a 50-minute flight and on the 13th flew from Etampes to Chevilly.

A few further details of construction may be given: the wings themselves and an elevator at the tail controlled the rate of ascent and descent, while a rudder was also fitted at the tail. The steering lever, working on a universally jointed shaft--forerunner of the modern joystick--controlled both the rudder and the wings, while a pedal actuated the elevator. The engine drove a two-bladed tractor screw of 6 feet 7 inches diameter, and the angle of incidence of the wings was 20 degrees. Timed at Issy, the speed of the machine was given as 36 miles an hour, and as Bleriot accomplished the Channel flight of 20 miles in 37 minutes, he probably had a slight following wind.

The Daily Mail had offered a prize of L1,000 for the first Cross-Channel flight, and Hubert Latham set his mind on winning it. He put up a shelter on the French coast at Sangatte, half-way between Calais and Cape Blanc Nez. From here he made his first attempt to fly to England on Monday the 19th of July. He soared to a fair height, circling, and reached an estimated height of about 900 feet as he came over the water with every appearance of capturing the Cross-Channel prize. The luck which dogged his career throughout was against him, for, after he had covered some 8 miles, his engine stopped and he came down to the water in a series of long glides. It was discovered afterward that a small piece of wire had worked its way into a vital part of the engine to rob Latham of the honour he coveted. The tug that came to his rescue found him seated on the fuselage of his Antoinette, smoking a cigarette and waiting for a boat to take him to the tug. It may be remarked that Latham merely assumed his Antoinette would float in case he failed to make the English coast; he had no actual proof.

Bleriot immediately entered his machine for the prize and took up his quarters at Barraques. On Sunday, July 25th, 1909, shortly after 4 a.m., Bleriot had his machine taken out from its shelter and prepared for flight. He had been recently injured in a petrol explosion and hobbled out on crutches to make his cross-Channel attempt; he made two great circles in the air to try the machine, and then alighted. 'In ten minutes I start for England,' he declared, and at 4.35 the motor was started up. After a run of 100 yards, the machine rose in the air and got a height of about 100 feet over the land, then wheeling sharply seaward and heading for Dover.

Bleriot had no means of telling direction, and any change of wind might have driven him out over the North Sea, to be lost, as were Cecil Grace and Hamel later on. Luck was with him, however, and at 5.12 a.m. of that July Sunday, he made his landing in the North Fall meadow, just behind Dover Castle. Twenty minutes out from the French coast, he lost sight of the destroyer which was patrolling the Channel, and at the same time he was out of sight of land without compass or any other means of ascertaining his direction. Sighting the English coast, he found that he had gone too far to the east, for the wind increased in strength throughout the flight, this to such an extent as almost to turn the machine round when he came over English soil. Profiting by Latham's experience, Bleriot had fitted an inflated rubber cylinder a foot in diameter by 5 feet in length along the middle of his fuselage, to render floating a certainty in case he had to alight on the water.

Latham in his camp at Sangatte had been allowed to sleep through the calm of the early morning through a mistake on the part of a friend, and when his machine was turned out--in order that he might emulate Bleriot, although he no longer hoped to make the first flight, it took so long to get the machine ready and dragged up to its starting-point that there was a 25 mile an hour wind by the time everything was in readiness. Latham was anxious to make the start in spite of the wind, but the Directors of the Antoinette Company refused permission. It was not until two days later that the weather again became favourable, and then with a fresh machine, since the one on which he made his first attempt had been very badly damaged in being towed ashore, he made a circular trial flight of about 5 miles. In landing from this, a side gust of wind drove the nose of the machine against a small hillock, damaging both propeller blades and chassis, and it was not until evening that the damage was repaired.

French torpedo boats were set to mark the route, and Latham set out on his second attempt at six o'clock. Flying at a height of 200 feet, he headed over the torpedo boats for Dover and seemed certain of making the English coast, but a mile and a half out from Dover his engine failed him again, and he dropped to the water to be picked up by the steam pinnace of an English warship and put aboard the French destroyer Escopette.

There is little to choose between the two aviators for courage in attempting what would have been considered a foolhardy feat a year or two before. Bleriot's state, with an abscess in the burnt foot which had to control the elevator of his machine, renders his success all the more remarkable. His machine was exhibited in London for a time, and was afterwards placed in the Conservatoire des Arts et Metiers, while a memorial in stone, copying his monoplane in form, was let into the turf at the point where he landed.

The second Channel crossing was not made until 1910, a year of new records. The altitude record had been lifted to over 10,000 feet, the duration record to 8 hours 12 minutes, and the distance for a single flight to 365 miles, while a speed of over 65 miles an hour had been achieved, when Jacques de Lesseps, son of the famous engineer of Suez Canal and Panama fame, crossed from France to England on a Bleriot monoplane. By this time flying had dropped so far from the marvellous that this second conquest of the Channel aroused but slight public interest in comparison with Bleriot's feat.

The total weight of Bleriot's machine in Cross Channel trim was 660 lbs., including the pilot and sufficient petrol for a three hours' run; at a speed of 37 miles an hour, it was capable of carrying about 5 lbs. per square foot of lifting surface. It was the three-cylinder 25 horse-power Anzani motor which drove the machine for the flight. Shortly after the flight had been accomplished, it was announced that the Bleriot firm would construct similar machines for sale at L400 apiece--a good commentary on the prices of those days.

On June the 2nd, 1910, the third Channel crossing was made by C. S. Rolls, who flew from Dover, got himself officially observed over French soil at Barraques, and then flew back without landing. He was the first to cross from the British side of the Channel and also was the first aviator who made the double journey. By that time, however, distance flights had so far increased as to reduce the value of the feat, and thenceforth the Channel crossing was no exceptional matter. The honour, second only to that of the Wright Brothers, remains with Bleriot.

XVI. LONDON TO MANCHESTER

The last of the great contests to arouse public enthusiasm was the London to Manchester Flight of 1910. As far back as 1906, the Daily Mail had offered a prize of L10,000 to the first aviator who should accomplish this journey, and, for a long time, the offer was regarded as a perfectly safe one for any person or paper to make--it brought forth far more ridicule than belief. Punch offered a similar sum to the first man who should swim the Atlantic and also for the first flight to Mars and back within a week, but in the spring of 1910 Claude Grahame White and Paulhan, the famous French pilot, entered for the 183 mile run on which the prize depended. Both these competitors flew the Farman biplane with the 50 horse-power Gnome motor as propulsive power. Grahame White surveyed the ground along the route, and the L. & N. W. Railway Company, at his request, whitewashed the sleepers for 100 yards on the north side of all junctions to give him his direction on the course. The machine was run out on to the starting ground at Park Royal and set going at 5.19 a.m. on April 23rd. After a run of 100 yards, the machine went up over Wormwood Scrubs on its journey to Normandy, near Hillmorten, which was the first arranged stopping place en route; Grahame White landed here in good trim at 7.20 a.m., having covered 75 miles and made a world's record cross country flight. At 8.15 he set off again to come down at Whittington, four miles short of Lichfield, at about 9.20, with his machine in good order except for a cracked landing skid. Twice, on this second stage of the journey, he had been caught by gusts of wind which turned the machine fully round toward London, and, when over a wood near Tamworth, the engine stopped through a defect in the balance springs of two exhaust valves; although it started up again after a 100 foot glide, it did not give enough power to give him safety in the gale he was facing. The rising wind kept him on the ground throughout the day, and, though he hoped for better weather, the gale kept up until the Sunday evening. The men in charge of the machine during its halt had attempted to hold the machine down instead of anchoring it with stakes and ropes, and, in consequence of this, the wind blew the machine over on its back, breaking the upper planes and the tail. Grahame White had to return to London, while the damaged machine was prepared for a second flight. The conditions of the competition enacted that the full journey should be completed within 24 hours, which made return to the starting ground inevitable.

Louis Paulhan, who had just arrived with his Farman machine, immediately got it unpacked and put together in order to be ready to make his attempt for the prize as soon as the weather conditions should admit. At 5.31 p.m., on April 27th, he went up from Hendon and had travelled 50 miles when Grahame White, informed of his rival's start, set out to overtake him. Before nightfall Paulhan landed at Lichfield, 117 miles from London, while Grahame White had to come down at Roden, only 60 miles out. The English aviator's chance was not so small as it seemed, for, as Latham had found in his cross-Channel attempts, engine failure was more the rule than the exception, and a very little thing might reverse the relative positions.

A special train accompanied Paulhan along the North-Western route, conveying Madame Paulhan, Henry Farman, and the mechanics who fitted the Farman biplane together. Paulhan himself, who had flown at a height of 1,000 feet, spent the night at Lichfield, starting again at 4.9 a.m. On the 28th, passing Stafford at 4.45, Crewe at 5.20, and landing at Burnage, near Didsbury, at 5.32, having had a clean run.

Meanwhile, Grahame White had made a most heroic attempt to beat his rival. An hour before dawn on the 28th, he went to the small field in which his machine had landed, and in the darkness managed to make an ascent from ground which made starting difficult even in daylight. Purely by instinct and his recollection of the aspect of things the night before, he had to clear telegraph wires and a railway bridge, neither of which he could possibly see at that hour. His engine, too, was faltering, and it was obvious to those who witnessed his start that its note was far from perfect.

At 3.50 he was over Nuneaton and making good progress; between Atherstone and Lichfield the wind caught him and the engine failed more and more, until at 4.13 in the morning he was forced to come to earth, having covered 6 miles less distance than in his first attempt. It was purely a case of engine failure, for, with full power, he would have passed over Paulhan just as the latter was preparing for the restart. Taking into consideration the two machines, there is little doubt that Grahame White showed the greater flying skill, although he lost the prize. After landing and hearing of Paulhan's victory, on which he wired congratulations, he made up his mind to fly to Manchester within the 24 hours. He started at 5 o'clock in the afternoon from Polesworth, his landing place, but was forced to land at 5.30 at Whittington, where he had landed on the previous Saturday. The wind, which had forced his descent, fell again and permitted of starting once more; on this third stage he reached Lichfield, only to make his final landing at 7.15 p.m., near the Trent Valley station. The defective running of the Gnome engine prevented his completing the course, and his Farman machine had to be brought back to London by rail.

The presentation of the prize to Paulhan was made the occasion for the announcement of a further competition, consisting of a 1,000 mile flight round a part of Great Britain. In this, nineteen competitors started, and only four finished; the end of the race was a great fight between Beaumont and Vedrines, both of whom scorned weather conditions in their determination to win. Beaumont made the distance in a flying time of 22 hours 28 minutes 19 seconds, and Vedrines covered the journey in a little over 23 1/2 hours. Valentine came third on a Deperdussin monoplane and S. F. Cody on his Cathedral biplane was fourth. This was in 1911, and by that time heavier-than-air flight had so far advanced that some pilots had had war experience in the Italian campaign in Tripoli, while long cross-country flights were an everyday event, and bad weather no longer counted.

XVII. A SUMMARY, TO 1911

There is so much overlapping in the crowded story of the first years of successful power-driven flight that at this point it is advisable to make a concise chronological survey of the chief events of the period of early development, although much of this is of necessity recapitulation. The story begins, of course, with Orville Wright's first flight of 852 feet at Kitty Hawk on December 19th, 1903. The next event of note was Wright's flight of 11.12 miles in 18 minutes 9 seconds at Dayton, Ohio, on September 26th, 1905, this being the first officially recorded flight. On October 4th of the same year, Wright flew 20.75 miles in 33 minutes 17 seconds, this being the first flight of over 20 miles ever made. Then on September 14th 1906, Alberto Santos-Dumont made a flight of eight seconds on the second heavier-than-air machine he had constructed. It was a big box-kite-like machine; this was the second power-driven aeroplane in Europe to fly, for although Santos-Dumont's first machine produced in 1905 was reckoned an unsuccessful design, it had actually got off the ground for brief periods. Louis Bleriot came into the ring on April 5th, 1907, with a first flight of 6 seconds on a Bleriot monoplane, his eighth but first successful construction.

Henry Farman made his first appearance in the history of aviation with a flight of 935 feet on a Voisin biplane on October 15th 1907. On October 25th, in a flight of 2,530 feet, he made the first recorded turn in the air, and on March 29th, 1908, carrying Leon Delagrange on a Voisin biplane, he made the first passenger flight. On April 10th of this year, Delagrange, in flying 1 1/2 miles, made the first flight in Europe exceeding a mile in distance. He improved on this by flying 10 1/2 miles at Milan on June 22nd, while on July 8th, at Turin, he took up Madame Peltier, the first woman to make an aeroplane flight.

Wilbur Wright, coming over to Europe, made his first appearance on the Continent with a flight of 1 3/4 minutes at Hunaudieres, France, on August 8th, 1908. On September 6th, at Chalons, he flew for 1 hour 4 minutes 26 seconds with a passenger, this being the first flight in which an hour in the air was exceeded with a passenger on board.

On September 12th 1908, Orville Wright, flying at Fort Meyer, U.S.A., with Lieut. Selfridge as passenger, crashed his machine, suffering severe injuries, while Selfridge was killed. This was the first aeroplane fatality. On October 30th, 1908, Farman made the first cross-country flight, covering the distance of 17 miles between Bouy and Rheims. The next day, Louis Bleriot, in flying from Toury to Artenay, made two landings en route, this being the first cross-country flight with landings. On the last day of the year, Wilbur Wright won the Michelin Cup at Auvours with a flight of 90 miles, which, lasting 2 hours 20 minutes 23 seconds, exceeded 2 hours in the air for the first time.

On January 2nd, 1909, S. F. Cody opened the New Year by making the first observed flight at Farnborough on a British Army aeroplane. It was not until July 18th of 1909 that the first European height record deserving of mention was put up by Paulhan, who achieved a height of 450 feet on a Voisin biplane. This preceded Latham's first attempt to fly the Channel by two days, and five days later, on the 25th of the month, Bleriot made the first Channel crossing. The Rheims Meeting followed on August 22nd, and it was a great day for aviation when nine machines were seen in the air at once. It was here that Farman, with a 118 mile flight, first exceeded the hundred miles, and Latham raised the height record officially to 500 feet, though actually he claimed to have reached 1,200 feet. On September 8th, Cody, flying from Aldershot, made a 40 mile journey, setting up a new cross-country record. On October 19th the Comte de Lambert flew from Juvisy to Paris, rounded the Eiffel Tower and flew back. J. T. C. Moore-Brabazon made the first circular mile flight by a British aviator on an all-British machine in Great Britain, on October 30th, flying a Short biplane with a Green engine. Paulhan, flying at Brooklands on November 2nd, accomplished 96 miles in 2 hours 48 minutes, creating a British distance record; on the following day, Henry Farman made a flight of 150 miles in 4 hours 22 minutes at Mourmelon, and on the 5th of the month, Paulhan, flying a Farman biplane, made a world's height record of 977 feet. This, however, was not to stand long, for Latham got up to 1,560 feet on an Antoinette at Mourmelon on December 1st. December 31st witnessed the first flight in Ireland, made by H. Ferguson on a monoplane which he himself had constructed at Downshire Park, Lisburn.

These, thus briefly summarised, are the principal events up to the end of 1909. 1910 opened with tragedy, for on January 4th Leon Delagrange, one of the greatest pilots of his time, was killed while flying at Pau. The machine was the Bleriot XI which Delagrange had used at the Doncaster meeting, and to which Delagrange had fitted a 50 horse-power Gnome engine, increasing the speed of the machine from its original 30 to 45 miles per hour. With the Rotary Gnome engine there was of necessity a certain gyroscopic effect, the strain of which proved too much for the machine. Delagrange had come to assist in the inauguration of the Croix d'Hins aerodrome, and had twice lapped the course at a height of about 60 feet. At the beginning of the third lap, the strain of the Gnome engine became too great for the machine; one wing collapsed as if the stay wires had broken, and the whole machine turned over and fell, killing Delagrange.

On January 7th Latham, flying at Mourmelon, first made the vertical kilometre and dedicated the record to Delagrange, this being the day of his friend's funeral. The record was thoroughly authenticated by a large registering barometer which Latham carried, certified by the officials of the French Aero Club. Three days later Paulhan, who was at Los Angeles, California, raised the height record to 4,146 feet.

On January 25th the Brussels Exhibition opened, when the Antoinette monoplane, the Gaffaux and Hanriot monoplanes, together with the d'Hespel aeroplane, were shown; there were also the dirigible Belgica and a number of interesting aero engines, including a German airship engine and a four-cylinder 50 horse-power Miesse, this last air-cooled by means of 22 fans driving a current of air through air jackets surrounding fluted cylinders.

On April 2nd Hubert Le Blon, flying a Bleriot with an Anzani engine, was killed while flying over the water. His machine was flying quite steadily, when it suddenly heeled over and came down sideways into the sea; the motor continued running for some seconds and the whole machine was drawn under water. When boats reached the spot, Le Blon was found lying back in the driving seat floating just below the surface. He had done good flying at Doncaster, and at Heliopolis had broken the world's speed records for 5 and 10 kilometres. The accident was attributed to fracture of one of the wing stay wires when running into a gust of wind.

The next notable event was Paulhan's London-Manchester flight, of which full details have already been given. In May Captain Bertram Dickson, flying at the Tours meeting, beat all the Continental fliers whom he encountered, including Chavez, the Peruvian, who later made the first crossing of the Alps. Dickson was the first British winner of international aviation prizes.

C. S. Rolls, of whom full details have already been given, was killed at Bournemouth on July 12th, being the first British aviator of note to be killed in an aeroplane accident. His return trip across the Channel had taken place on June 2nd. Chavez, who was rapidly leaping into fame, as a pilot, raised the British height record to 5,750 feet while flying at Blackpool on August 3rd. On the 11th of that month, Armstrong Drexel, flying a Bleriot, made a world's height record of 6,745 feet.

It was in 1910 that the British War office first began fully to realise that there might be military possibilities in heavier-than-air flying. C. S. Rolls had placed a Wright biplane at the disposal of the military authorities, and Cody, as already recorded, had been experimenting with a biplane type of his own for some long period. Such development as was achieved was mainly due to the enterprise and energy of Colonel J. E. Capper, C.B., appointed to the superintendency of the Balloon Factory and Balloon School at Farnborough in 1906. Colonel Capper's retirement in 1910 brought (then) Mr Mervyn O'Gorman to command, and by that time the series of successes of the Cody biplane, together with the proved efficiency of the aeroplane in various civilian meetings, had convinced the British military authorities that the mastery of the air did not lie altogether with dirigible airships, and it may be said that in 1910 the British War office first began seriously to consider the possibilities of the aeroplane, though two years more were to elapse before the formation of the Royal Flying Corps marked full realisation of its value.

A triumph and a tragedy were combined in September of 1910. On the 23rd of the month, Georges Chavez set out to fly across the Alps on a Bleriot monoplane. Prizes had been offered by the Milan Aviation Committee for a flight from Brigue in Switzerland over the Simplon Pass to Milan, a distance of 94 miles with a minimum height of 6,600 feet above sea level. Chavez started at 1.30 p.m. On the 23rd, and 41 minutes later he reached Domodossola, 25 miles distant. Here he descended, numbed with the cold of the journey; it was said that the wings of his machine collapsed when about 30 feet from the ground, but however this may have been, he smashed the machine on landing, and broke both legs, in addition to sustaining other serious injuries. He lay in hospital until the 27th September, when he died, having given his life to the conquest of the Alps. His death in the moment of success was as great a tragedy as were those of Pilcher and Lilienthal.

The day after Chavez's death, Maurice Tabuteau flew across the Pyrenees, landing in the square at Biarritz. On December 30th, Tabuteau made a flight of 365 miles in 7 hours 48 minutes. Farman, on December 18th, had flown for over 8 hours, but his total distance was only 282 miles. The autumn of this year was also noteworthy for the fact that aeroplanes were first successfully used in the French Military Manoeuvres. The British War Office, by the end of the year, had bought two machines, a military type Farman and a Paulhan, ignoring British experimenters and aeroplane builders of proved reliability. These machines, added to an old Bleriot two-seater, appear to have constituted the British aeroplane fleet of the period.

There were by this time three main centres of aviation in England, apart from Cody, alone on Laffan's Plain. These three were Brooklands, Hendon, and the Isle of Sheppey, and of the three Brooklands was chief. Here such men as Graham Gilmour, Rippen, Leake, Wickham, and Thomas persistently experimented. Hendon had its own little group, and Shellbeach, Isle of Sheppey, held such giants of those days as C. S. Rolls and Moore Brabazon, together with Cecil Grace and Rawlinson. One or other, and sometimes all of these were deserted on the occasion of some meeting or other, but they were the points where the spade work was done, Brooklands taking chief place. 'If you want the early history of flying in England, it is there,' one of the early school remarked, pointing over toward Brooklands course.

1911 inaugurated a new series of records of varying character. On the 17th January, E. B. Ely, an American, flew from the shore of San Francisco to the U.S. cruiser Pennsylvania, landing on the cruiser, and then flew back to the shore. The British military designing of aeroplanes had been taken up at Farnborough by G. H. de Havilland, who by the end of January was flying a machine of his own design, when he narrowly escaped becoming a casualty through collision with an obstacle on the ground, which swept the undercarriage from his machine.

A list of certified pilots of the countries of the world was issued early in 1911, showing certificates granted up to the end of 1910. France led the way easily with 353 pilots; England came next with 57, and Germany next with 46; Italy owned 32, Belgium 27, America 26, and Austria 19; Holland and Switzerland had 6 aviators apiece, while Denmark followed with 3, Spain with 2, and Sweden with 1. The first certificate in England was that of J. T. C. Moore-Brabazon, while Louis Bleriot was first on the French list and Glenn Curtiss, first holder of an American certificate, also held the second French brevet.

On the 7th March, Eugene Renaux won the Michelin Grand Prize by flying from the French Aero Club ground at St Cloud and landing on the Puy de Dome. The landing, which was one of the conditions of the prize, was one of the most dangerous conditions ever attached to a competition; it involved dropping on to a little plateau 150 yards square, with a possibility of either smashing the machine against the face of the mountain, or diving over the edge of the plateau into the gulf beneath. The length of the journey was slightly over 200 miles and the height of the landing point 1,465 metres, or roughly 4,500 feet above sea-level. Renaux carried a passenger, Doctor Senoucque, a member of Charcot's South Polar Expedition.

The 1911 Aero Exhibition held at Olympia bore witness to the enormous strides made in construction, more especially by British designers, between 1908 and the opening of the Show. The Bristol Firm showed three machines, including a military biplane, and the first British built biplane with tractor screw. The Cody biplane, with its enormous size rendering it a prominent feature of the show, was exhibited. Its designer anticipated later engines by expressing his desire for a motor of 150 horse-power, which in his opinion was necessary to get the best results from the machine. The then famous Dunne monoplane was exhibited at this show, its planes being V-shaped in plan, with apex leading. It embodied the results of very lengthy experiments carried out both with gliders and power-driven machines by Colonel Capper, Lieut. Gibbs, and Lieut. Dunne, and constituted the longest step so far taken in the direction of inherent stability.

Such forerunners of the notable planes of the war period as the Martin Handasyde, the Nieuport, Sopwith, Bristol, and Farman machines, were features of the show; the Handley-Page monoplane, with a span of 32 feet over all, a length of 22 feet, and a weight of 422 lbs., bore no relation at all to the twin-engined giant which later made this firm famous. In the matter of engines, the principal survivals to the present day, of which this show held specimens, were the Gnome, Green, Renault air-cooled, Mercedes four-cylinder dirigible engine of 115 horse-power, and 120 horsepower Wolseley of eight cylinders for use with dirigibles.

On April 12th, of 1911, Paprier, instructor at the Bleriot school at Hendon, made the first non-stop flight between London and Paris. He left the aerodrome at 1.37 p.m., and arrived at Issy-les-Moulineaux at 5.33 p.m., thus travelling 250 miles in a little under 4 hours. He followed the railway route practically throughout, crossing from Dover to nearly opposite Calais, keeping along the coast to Boulogne, and then following the Nord Railway to Amiens, Beauvais, and finally Paris.

In May, the Paris-Madrid race took place; Vedrines, flying a Morane biplane, carried off the prize by first completing the distance of 732 miles. The Paris-Rome race of 916 miles was won in the same month by Beaumont, flying a Bleriot monoplane. In July, Koenig won the German National Circuit race of 1,168 miles on an Albatross biplane. This was practically simultaneous with the Circuit of Britain won by Beaumont, who covered 1,010 miles on a Bleriot monoplane, having already won the Paris-Brussels-London-Paris Circuit of 1,080 miles, this also on a Bleriot. It was in August that a new world's height record of 11,152 feet was set up by Captain Felix at Etampes, while on the 7th of the month Renaux flew nearly 600 miles on a Maurice Farman machine in 12 hours. Cody and Valentine were keeping interest alive in the Circuit of Britain race, although this had long been won, by determinedly plodding on at finishing the course.

On September 9th, the first aerial post was tried between Hendon and Windsor, as an experiment in sending mails by aeroplane. Gustave Hamel flew from Hendon to Windsor and back in a strong wind. A few days later, Hamel went on strike, refusing to carry further mails unless the promoters of the Aerial Postal Service agreed to pay compensation to Hubert, who fractured both his legs on the 11th of the month while engaged in aero postal work. The strike ended on September 25th, when Hamel resumed mail-carrying in consequence of the capitulation of the Postmaster-General, who agreed to set aside L500 as compensation to Hubert.

September also witnessed the completion in America of a flight across the Continent, a distance of 2,600 miles. The only competitor who completed the full distance was C. P. Rogers, who was disqualified through failing to comply with the time limit. Rogers needed so many replacements to his machine on the journey that, expressing it in American fashion, he arrived with practically a dfferent aeroplane from that with which he started.

With regard to the aerial postal service, analysis of the matter carried and the cost of the service seemed to show that with a special charge of one shilling for letters and sixpence for post cards, the revenue just balanced the expenditure. It was not possible to keep to the time-table as, although the trials were made in the most favourable season of the year, aviation was not sufficiently advanced to admit of facing all weathers and complying with time-table regulations.

French military aeroplane trials took place at Rheims in October, the noteworthy machines being Antoinette, Farman, Nieuport, and Deperdussin. The tests showed the Nieuport monoplane with Gnome motor as first in position; the Breguet biplane was second, and the Deperdussin monoplanes third. The first five machines in order of merit were all engined with the Gnome motor.

The records quoted for 1911 form the best evidence that can be given of advance in design and performance during the year. It will be seen that the days of the giants were over; design was becoming more and more standardised and aviation not so much a matter of individual courage and even daring, as of the reliability of the machine and its engine. This was the first year in which the twin-engined aeroplane made its appearance, and it was the year, too, in which flying may be said to have grown so common that the 'meetings' which began with Rheims were hardly worth holding, owing to the fact that increase in height and distance flown rendered it no longer necessary for a would-be spectator of a flight to pay half a crown and enter an enclosure. Henceforth, flying as a spectacle was very little to be considered; its commercial aspects were talked of, and to a very slight degree exploited, but, more and more, the fact that the aeroplane was primarily an engine of war, and the growing German menace against the peace of the world combined to point the way of speediest development, and the arrangements for the British Military Trials to be held in August, 1912, showed that even the British War office was waking up to the potentialities of this new engine of war.

XVIII. A SUMMARY, TO 1914

Consideration of the events in the years immediately preceding the War must be limited to as brief a summary as possible, this not only because the full history of flying achievements is beyond the compass of any single book, but also because, viewing the matter in perspective, the years 1903-1911 show up as far more important as regards both design and performance. From 1912 to August of 1914, the development of aeronautics was hindered by the fact that it had not progressed far enough to form a real commercial asset in any country. The meetings which drew vast concourses of people to such places as Rheims and Bournemouth may have been financial successes at first, but, as flying grew more common and distances and heights extended, a great many people found it other than worth while to pay for admission to an aerodrome. The business of taking up passengers for pleasure flights was not financially successful, and, although schemes for commercial routes were talked of, the aeroplane was not sufficiently advanced to warrant the investment of hard cash in any of these projects. There was a deadlock; further development was necessary in order to secure financial aid, and at the same time financial aid was necessary in order to secure further development. Consequently, neither was forthcoming.

This is viewing the matter in a broad and general sense; there were firms, especially in France, but also in England and America, which looked confidently for the great days of flying to arrive, and regarded their sunk capital as investment which would eventually bring its due return. But when one looks back on those years, the firms in question stand out as exceptions to the general run of people, who regarded aeronautics as something extremely scientific, exceedingly dangerous, and very expensive. The very fame that was attained by such pilots as became casualties conduced to the advertisement of every death, and the dangers attendant on the use of heavier-than-air machines became greatly exaggerated; considering the matter as one of number of miles flown, even in the early days, flying exacted no more toll in human life than did railways or road motors in the early stages of their development. But to take one instance, when C. S. Rolls was killed at Bournemouth by reason of a faulty tail-plane, the fact was shouted to the whole world with almost as much vehemence as characterised the announcement of the Titanic sinking in mid-Atlantic.

Even in 1911 the deadlock was apparent; meetings were falling off in attendance, and consequently in financial benefit to the promoters; there remained, however, the knowledge--for it was proved past question--that the aeroplane in its then stage of development was a necessity to every army of the world. France had shown this by the more than interest taken by the French Government in what had developed into an Air Section of the French army; Germany, of course, was hypnotised by Count Zeppelin and his dirigibles, to say nothing of the Parsevals which had been proved useful military accessories; in spite of this, it was realised in Germany that the aeroplane also had its place in military affairs. England came into the field with the military aeroplane trials of August 1st to 15th, 1912, barely two months after the founding of the Royal Flying Corps.

When the R.F.C. was founded--and in fact up to two years after its founding--in no country were the full military potentialities of the aeroplane realised; it was regarded as an accessory to cavalry for scouting more than as an independent arm; the possibilities of bombing were very vaguely considered, and the fact that it might be possible to shoot from an aeroplane was hardly considered at all. The conditions of the British Military Trials of 1912 gave to the War office the option of purchasing for L1,000 any machine that might be awarded a prize. Machines were required, among other things, to carry a useful load of 350 lbs. in addition to equipment, with fuel and oil for 4 1/2-hours; thus loaded, they were required to fly for 3 hours, attaining an altitude of 4,500 feet, maintaining a height of 1,500 feet for 1 hour, and climbing 1,000 feet from the ground at a rate of 200 feet per minute, 'although 300 feet per minute is desirable.' They had to attain a speed of not less than 55 miles per hour in a calm, and be able to plane down to the ground in a calm from not more than 1,000 feet with engine stopped, traversing 6,000 feet horizontal distance. For those days, the landing demands were rather exacting; the machine should be able to rise without damage from long grass, clover, or harrowed land, in 100 yards in a calm, and should be able to land without damage on any cultivated ground, including rough ploughed land, and, when landing on smooth turf in a calm, be able to pull up within 75 yards of the point of first touching the ground. It was required that pilot and observer should have as open a view as possible to front and flanks, and they should be so shielded from the wind as to be able to communicate with each other. These are the main provisions out of the set of conditions laid down for competitors, but a considerable amount of leniency was shown by the authorities in the competition, who obviously wished to try out every machine entered and see what were its capabilities.

The beginning of the competition consisted in assembling the machines against time from road trim to flying trim. Cody's machine, which was the only one to be delivered by air, took 1 hour and 35 minutes to assemble; the best assembling time was that of the Avro, which was got into flying trim in 14 minutes 30 seconds. This machine came to grief with Lieut. Parke as pilot, on the 7th, through landing at very high speed on very bad ground; a securing wire of the under-carriage broke in the landing, throwing the machine forward on to its nose and then over on its back. Parke was uninjured, fortunately; the damaged machine was sent off to Manchester for repair and was back again on the 16th of August.

It is to be noted that by this time the Royal Aircraft Factory was building aeroplanes of the B.E. and F.E. types, but at the same time it is also to be noted that British military interest in engines was not sufficient to bring them up to the high level attained by the planes, and it is notorious that even the outbreak of war found England incapable of providing a really satisfactory aero engine. In the 1912 Trials, the only machines which actually completed all their tests were the Cody biplane, the French Deperdussin, the Hanriot, two Bleriots and a Maurice Farman. The first prize of L4,000, open to all the world, went to F. S. Cody's British-built biplane, which complied with all the conditions of the competition and well earned its official acknowledgment of supremacy. The machine climbed at 280 feet per minute and reached a height of 5,000 feet, while in the landing test, in spite of its great weight and bulk, it pulled up on grass in 56 yards. The total weight was 2,690 lbs. when fully loaded, and the total area of supporting surface was 500 square feet; the motive power was supplied by a six-cylinder 120 horsepower Austro-Daimler engine. The second prize was taken by A. Deperdussin for the French-built Deperdussin monoplane. Cody carried off the only prize awarded for a British-built plane, this being the sum of L1,000, and consolation prizes of L500 each were awarded to the British Deperdussin Company and The British and Colonial Aeroplane Company, this latter soon to become famous as makers of the Bristol aeroplane, of which the war honours are still fresh in men's minds.

While these trials were in progress Audemars accomplished the first flight between Paris and Berlin, setting out from Issy early in the morning of August 18th, landing at Rheims to refill his tanks within an hour and a half, and then coming into bad weather which forced him to land successively at Mezieres, Laroche, Bochum, and finally nearly Gersenkirchen, where, owing to a leaky petrol tank, the attempt to win the prize offered for the first flight between the two capitals had to be abandoned after 300 miles had been covered, as the time limit was definitely exceeded. Audemars determined to get through to Berlin, and set off at 5 in the morning of the 19th, only to be brought down by fog; starting off again at 9.15 he landed at Hanover, was off again at 1.35, and reached the Johannisthal aerodrome in the suburbs of Berlin at 6.48 that evening.

As early as 1910 the British Government possessed some ten aeroplanes, and in 1911 the force developed into the Army Air Battalion, with the aeroplanes under the control of Major J. H. Fulton, R.F.A. Toward the end of 1911 the Air Battalion was handed over to (then) Brig.-Gen. D. Henderson, Director of Military Training. On June 6th, 1912, the Royal Flying Corps was established with a military wing under Major F. H. Sykes and a naval wing under Commander C. R. Samson. A joint Naval and Military Flying School was established at Upavon with Captain Godfrey M. Paine, R.N., as Commandant and Major Hugh Trenchard as Assistant Commandant. The Royal Aircraft Factory brought out the B.E. and F.E. types of biplane, admittedly superior to any other British design of the period, and an Aircraft Inspection Department was formed under Major J. H. Fulton. The military wing of the R.F.C. was equipped almost entirely with machines of Royal Aircraft Factory design, but the Navy preferred to develop British private enterprise by buying machines from private firms. On July 1st, 1914 the establishment of the Royal Naval Air Service marked the definite separation of the military and naval sides of British aviation, but the Central Flying School at Upavon continued to train pilots for both services.

It is difficult at this length of time, so far as the military wing was concerned, to do full justice to the spade work done by Major-General Sir David Henderson in the early days. Just before war broke out, British military air strength consisted officially of eight squadrons, each of 12 machines and 13 in reserve, with the necessary complement of road transport. As a matter of fact, there were three complete squadrons and a part of a fourth which constituted the force sent to France at the outbreak of war. The value of General Henderson's work lies in the fact that, in spite of official stinginess and meagre supplies of every kind, he built up a skeleton organisation so elastic and so well thought out that it conformed to war requirements as well as even the German plans fitted in with their aerial needs. On the 4th of August, 1914, the nominal British air strength of the military wing was 179 machines. Of these, 82 machines proceeded to France, landing at Amiens and flying to Maubeuge to play their part in the great retreat with the British Expeditionary Force, in which they suffered heavy casualties both in personnel and machines. The history of their exploits, however, belongs to the War period.

The development of the aeroplane between 1912 and 1914 can be judged by comparison of the requirements of the British War Office in 1912 with those laid down in an official memorandum issued by the War Office in February, 1914. This latter called for a light scout aeroplane, a single-seater, with fuel capacity to admit of 300 miles range and a speed range of from 50 to 85 miles per hour. It had to be able to climb 3,500 feet in five minutes, and the engine had to be so constructed that the pilot could start it without assistance. At the same time, a heavier type of machine for reconnaissance work was called for, carrying fuel for a 200 mile flight with a speed range of between 35 and 60 miles per hour, carrying both pilot and observer. It was to be equipped with a wireless telegraphy set, and be capable of landing over a 30 foot vertical obstacle and coming to rest within a hundred yards' distance from the obstacle in a wind of not more than 15 miles per hour. A third requirement was a heavy type of fighting aeroplane accommodating pilot and gunner with machine gun and ammunition, having a speed range of between 45 and 75 miles per hour and capable of climbing 3,500 feet in 8 minutes. It was required to carry fuel for a 300 mile flight and to give the gunner a clear field of fire in every direction up to 30 degrees on each side of the line of flight. Comparison of these specifications with those of the 1912 trials will show that although fighting, scouting, and reconnaissance types had been defined, the development of performance compared with the marvellous development of the earlier years of achieved flight was small.

Yet the records of those years show that here and there an outstanding design was capable of great things. On the 9th September, 1912, Vedrines, flying a Deperdussin monoplane at Chicago, attained a speed of 105 miles an hour. On August 12th, G. de Havilland took a passenger to a height of 10,560 feet over Salisbury Plain, flying a B.E. biplane with a 70 horse-power Renault engine. The work of de Havilland may be said to have been the principal influence in British military aeroplane design, and there is no doubt that his genius was in great measure responsible for the excellence of the early B.E. and F.E. types.

On the 31st May, 1913, H. G. Hawker, flying at Brooklands, reached a height of 11,450 feet on a Sopwith biplane engined with an 80 horse-power Gnome engine. On June 16th, with the same type of machine and engine, he achieved 12,900 feet. On the 2nd October, in the same year, a Grahame White biplane with 120 horse-power Austro-Daimler engine, piloted by Louis Noel, made a flight of just under 20 minutes carrying 9 passengers. In France a Nieuport monoplane piloted by G. Legagneaux attained a height of 6,120 metres, or just over 20,070 feet, this being the world's height record. It is worthy of note that of the world's aviation records as passed by the International Aeronautical Federation up to June 30th, 1914, only one, that of Noel, is credited to Great Britain.

Just as records were made abroad, with one exception, so were the really efficient engines. In England there was the Green engine, but the outbreak of war found the Royal Flying Corps with 80 horse-power Gnomes, 70 horse-power Renaults, and one or two Antoinette motors, but not one British, while the Royal Naval Air Service had got 20 machines with engines of similar origin, mainly land planes in which the wheeled undercarriages had been replaced by floats. France led in development, and there is no doubt that at the outbreak of war, the French military aeroplane service was the best in the world. It was mainly composed of Maurice Farman two-seater biplanes and Bleriot monoplanes--the latter type banned for a period on account of a number of serious accidents that took place in 1912.

America had its Army Aviation School, and employed Burgess-Wright and Curtiss machines for the most part. In the pre-war years, once the Wright Brothers had accomplished their task, America's chief accomplishment consisted in the development of the 'Flying Boat,' alternatively named with characteristic American clumsiness, 'The Hydro-Aeroplane.' In February of 1911, Glenn Curtiss attached a float to a machine similar to that with which he won the first Gordon-Bennett

Air Contest and made his first flying boat experiment. From this beginning he developed the boat form of body which obviated the use and troubles of floats--his hydroplane became its own float.

Mainly owing to greater engine reliability the duration records steadily increased. By September of 1912 Fourny, on a Maurice Farman biplane, was able to accomplish a distance of 628 miles without a landing, remaining in the air for 13 hours 17 minutes and just over 57 seconds. By 1914 this was raised by the German aviator, Landemann, to 21 hours 48 3/4 seconds. The nature of this last record shows that the factors in such a record had become mere engine endurance, fuel capacity, and capacity of the pilot to withstand air conditions for a prolonged period, rather than any exceptional flying skill.

Let these years be judged by the records they produced, and even then they are rather dull. The glory of achievement such as characterised the work of the Wright Brothers, of Bleriot, and of the giants of the early days, had passed; the splendid courage, the patriotism and devotion of the pilots of the War period had not yet come to being. There was progress, past question, but it was mechanical, hardly ever inspired. The study of climatic conditions was definitely begun and aeronautical meteorology came to being, while another development already noted was the fitting of wireless telegraphy to heavier-than-air machines, as instanced in the British War office specification of February, 1914. These, however, were inevitable; it remained for the War to force development beyond the inevitable, producing in five years that which under normal circumstances might easily have occupied fifty--the aeroplane of to-day; for, as already remarked, there was a deadlock, and any survey that may be made of the years 1912-1914, no matter how superficial, must take it into account with a view to retaining correct perspective in regard to the development of the aeroplane.

There is one story of 1914 that must be included, however briefly, in any record of aeronautical achievement, since it demonstrates past question that to Professor Langley really belongs the honour of having achieved a design which would ensure actual flight, although the series of accidents which attended his experiments gave to the Wright Brothers the honour of first leaving the earth and descending without accident in a power-driven heavier-than-air machine. In March, 1914, Glenn Curtiss was invited to send a flying boat to Washington for the celebration of 'Langley

Day,' when he remarked, 'I would like to put the Langley aeroplane itself in the air.' In consequence of this remark, Secretary Walcot of the Smithsonian Institution authorised Curtiss to re-canvas the original Langley aeroplane and launch it either under its own power or with a more recent engine and propeller. Curtiss completed this, and had the machine ready on the shores of Lake Keuka, Hammondsport, N.Y., by May. The main object of these renewed trials was to show whether the original Langley machine was capable of sustained free flight with a pilot, and a secondary object was to determine more fully the advantages of the tandem monoplane type; thus the aeroplane was first flown as nearly as possible in its original condition, and then with such modifications as seemed desirable. The only difference made for the first trials consisted in fitting floats with connecting trusses; the steel main frame, wings, rudders, engine, and propellers were substantially as they had been in 1903. The pilot had the same seat under the main frame and the same general system of control. He could raise or lower the craft by moving the rear rudder up and down; he could steer right or left by moving the vertical rudder. He had no ailerons nor wing-warping mechanism, but for lateral balance depended on the dihedral angle of the wings and upon suitable movements of his weight or of the vertical rudder.

After the adjustments for actual flight had been made in the Curtiss factory, according to the minute descriptions contained in the Langley Memoir on Mechanical Flight, the aeroplane was taken to the shore of Lake Keuka, beside the Curtiss hangars, and assembled for launching. On a clear morning (May 28th) and in a mild breeze, the craft was lifted on to the water by a dozen men and set going, with Mr Curtiss at the steering wheel, esconced in the little boat-shaped car under the forward part of the frame. The four-winged craft, pointed somewhat across the wind, went skimming over the waveless, then automatically headed into the wind, rose in level poise, soared gracefully for 150 feet, and landed softly on the water near the shore. Mr Curtiss asserted that he could have flown farther, but, being unused to the machine, imagined that the left wings had more resistance than the right. The truth is that the aeroplane was perfectly balanced in wing resistance, but turned on the water like a weather vane, owing to the lateral pressure on its big rear rudder. Hence in future experiments this rudder was made turnable about a vertical axis, as well as about the horizontal axis used by Langley. Henceforth the little vertical rudder under the frame was kept fixed and inactive.[7]

That the Langley aeroplane was subsequently fitted with an 80 horse-power Curtiss engine and successfully flown is of little interest in such a record as this, except for the fact that with the weight nearly doubled by the new engine and accessories the machine flew successfully, and demonstrated the perfection of Langley's design by standing the strain. The point that is of most importance is that the design itself proved a success and fully vindicated Langley's work. At the same time, it would be unjust to pass by the fact of the flight without according to Curtiss due recognition of the way in which he paid tribute to the genius of the pioneer by these experiments.

XIX. THE WAR PERIOD--I

Full record of aeronautical progress and of the accomplishments of pilots in the years of the War would demand not merely a volume, but a complete library, and even then it would be barely possible to pay full tribute to the heroism of pilots of the war period. There are names connected with that period of which the glory will not fade, names such as Bishop, Guynemer, Boelcke, Ball, Fonck, Immelmann, and many others that spring to mind as one recalls the 'Aces' of the period. In addition to the pilots, there is the stupendous development of the machines--stupendous when the length of the period in which it was achieved is considered.

The fact that Germany was best prepared in the matter of heavier-than-air service machines in spite of the German faith in the dirigible is one more item of evidence as to who forced hostilities. The Germans came into the field with well over 600 aeroplanes, mainly two-seaters of standardised design, and with factories back in the Fatherland turning out sufficient new machines to make good the losses. There were a few single-seater scouts built for speed, and the two-seater machines were all fitted with cameras and bomb-dropping gear. Manoeuvres had determined in the German mind what should be the uses of the air fleet; there was photography of fortifications and field works; signalling by Very lights; spotting for the guns, and scouting for news of enemy movements. The methodical German mind had

---
7 Smithsonian Publications No. 2329.

arranged all this beforehand, but had not allowed for the fact that opponents might take counter-measures which would upset the over-perfect mechanism of the air service just as effectually as the great march on Paris was countered by the genius of Joffre.

The French Air Force at the beginning of the War consisted of upwards of 600 machines. These, unlike the Germans, were not standardised, but were of many and diverse types. In order to get replacements quickly enough, the factories had to work on the designs they had, and thus for a long time after the outbreak of hostilities standardisation was an impossibility. The versatility of a Latin race in a measure compensated for this; from the outset, the Germans tried to overwhelm the French Air Force, but failed, since they had not the numerical superiority, nor--this equally a determining factor--the versatility and resource of the French pilots. They calculated on a 50 per cent superiority to ensure success; they needed more nearly 400 per cent, for the German fought to rule, avoiding risks whenever possible, and definitely instructed to save both machines and pilots wherever possible. French pilots, on the other hand, ran all the risks there were, got news of German movements, bombed the enemy, and rapidly worked up a very respectable antiaircraft force which, whatever it may have accomplished in the way of hitting German planes, got on the German pilots' nerves.

It has already been detailed how Britain sent over 82 planes as its contribution to the military aerial force of 1914. These consisted of Farman, Caudron, and Short biplanes, together with Bleriot, Deperdussin and Nieuport monoplanes, certain R.A.F. types, and other machines of which even the name barely survives--the resourceful Yankee entitles them 'orphans.' It is on record that the work of providing spares might have been rather complicated but for the fact that there were none.

There is no doubt that the Germans had made study of aerial military needs just as thoroughly as they had perfected their ground organisation. Thus there were 21 illuminated aircraft stations in Germany before the War, the most powerful being at Weimar, where a revolving electric flash of over 27 million candle-power was located. Practically all German aeroplane tests in the period immediately preceding the War were of a military nature, and quite a number of reliability tests were carried out just on the other side of the French frontier. Night flying and landing were standardised items in the German pilot's course of instruction while they were still experimental in other countries, and a system of signals was arranged which rendered the instructional course as perfect as might be.

The Belgian contribution consisted of about twenty machines fit for active service and another twenty which were more or less useful as training machines. The material was mainly French, and the Belgian pilots used it to good account until German numbers swamped them. France, and to a small extent England, kept Belgian aviators supplied with machines throughout the War.

The Italian Air Fleet was small, and consisted of French machines together with a percentage of planes of Italian origin, of which the design was very much a copy of French types. It was not until the War was nearing its end that the military and naval services relied more on the home product than on imports. This does not apply to engines, however, for the F.I.A.T. and S.C.A.T. were equal to practically any engine of Allied make, both in design and construction.

Russia spent vast sums in the provision of machines: the giant Sikorsky biplane, carrying four 100 horsepower Argus motors, was designed by a young Russian engineer in the latter part of 1913, and in its early trials it created a world's record by carrying seven passengers for 1 hour 54 minutes. Sikorsky also designed several smaller machines, tractor biplanes on the lines of the British B.E. type, which were very successful. These were the only home productions, and the imports consisted mainly of French aeroplanes by the hundred, which got as far as the docks and railway sidings and stayed there, while German influence and the corruption that ruined the Russian Army helped to lose the War. A few Russian aircraft factories were got into operation as hostilities proceeded, but their products were negligible, and it is not on record that Russia ever learned to manufacture a magneto.

The United States paid tribute to British efficiency by adopting the British system of training for its pilots; 500 American cadets were trained at the School of Military Aeronautics at oxford, in order to form a nucleus for the American aviation schools which were subsequently set up in the United States and in France. As regards production of craft, the designing of the Liberty engine and building of over 20,000 aeroplanes within a year proves that America is a manufacturing country, even under the strain of war.

There were three years of struggle for aerial supremacy, the combatants being England and France against Germany, and the contest was neck and neck all the way. Germany led at the outset with the standardised two-seater biplanes manned by pilots and observers, whose training was superior to that afforded by any other nation, while the machines themselves were better equipped and fitted with accessories. All the early German aeroplanes were designated Taube by the uninitiated, and were formed with swept-back, curved wings very much resembling the wings of a bird. These had obvious disadvantages, but the standardisation of design and mass production of the German factories kept them in the field for a considerable period, and they flew side by side with tractor biplanes of improved design. For a little time, the Fokker monoplane became a definite threat both to French and British machines. It was an improvement on the Morane French monoplane, and with a high-powered engine it climbed quickly and flew fast, doing a good deal of damage for a brief period of 1915. Allied design got ahead of it and finally drove it out of the air.

German equipment at the outset, which put the Allies at a disadvantage, included a hand-operated magneto engine-starter and a small independent screw which, mounted on one of the main planes, drove the dynamo used for the wireless set. Cameras were fitted on practically every machine; equipment included accurate compasses and pressure petrol gauges, speed and height recording instruments, bomb-dropping fittings and sectional radiators which facilitated repairs and gave maximum engine efficiency in spite of variations of temperature. As counter to these, the Allied pilots had resource amounting to impudence. In the early days they carried rifles and hand grenades and automatic pistols. They loaded their machines down, often at their own expense, with accessories and fittings until their aeroplanes earned their title of Christmas trees. They played with death in a way that shocked the average German pilot of the War's early stages, declining to fight according to rule and indulging in the individual duels of the air which the German hated. As Sir John French put it in one of his reports, they established a personal ascendancy over the enemy, and in this way compensated for their inferior material.

French diversity of design fitted in well with the initiative and resource displayed by the French pilots. The big Caudron type was the ideal bomber of the early days; Farman machines were excellent for reconnaissance and artillery spotting; the Bleriots proved excellent as fighting scouts and for aerial photography; the Nieuports made good fighters, as did the Spads, both being very fast craft, as were the

Morane-Saulnier monoplanes, while the big Voisin biplanes rivalled the Caudron machines as bombers.

The day of the Fokker ended when the British B.E.2.C. aeroplane came to France in good quantities, and the F.E. type, together with the De Havilland machines, rendered British aerial superiority a certainty. Germany's best reply--this was about 1916--was the Albatross biplane, which was used by Captain Baron von Richthofen for his famous travelling circus, manned by German star pilots and sent to various parts of the line to hearten up German troops and aviators after any specially bad strafe. Then there were the Aviatik biplane and the Halberstadt fighting scout, a cleanly built and very fast machine with a powerful engine with which Germany tried to win back superiority in the third year of the War, but Allied design kept about three months ahead of that of the enemy, once the Fokker had been mastered, and the race went on. Spads and Bristol fighters, Sopwith scouts and F.E.'s played their part in the race, and design was still advancing when peace came.

The giant twin-engined Handley-Page bomber was tried out, proved efficient, and justly considered better than anything of its kind that had previously taken the field. Immediately after the conclusion of its trials, a specimen of the type was delivered intact at Lille for the Germans to copy, the innocent pilot responsible for the delivery doing some great disservice to his own cause. The Gotha Wagon-Fabrik Firm immediately set to work and copied the Handley-Page design, producing the great Gotha bombing machine which was used in all the later raids on England as well as for night work over the Allied lines.

How the War advanced design may be judged by comparison of the military requirements given for the British Military Trials of 1912, with performances of 1916 and 1917, when the speed of the faster machines had increased to over 150 miles an hour and Allied machines engaged enemy aircraft at heights ranging up to 22,000 feet. All pre-war records of endurance, speed, and climb went by the board, as the race for aerial superiority went on.

Bombing brought to being a number of crude devices in the first year of the War. Allied pilots of the very early days carried up bombs packed in a small box and threw them over by hand, while, a little later, the bombs were strung like apples on wings and undercarriage, so that the pilot who did not get rid of his load before landing risked an explosion. Then came a properly designed carrying apparatus, crude but fairly efficient, and with 1916 development had proceeded as far as the proper bomb-racks with releasing gear.

Reconnaissance work developed, so that fighting machines went as escort to observing squadrons and scouting operations were undertaken up to 100 miles behind the enemy lines; out of this grew the art of camouflage, when ammunition dumps were painted to resemble herds of cows, guns were screened by foliage or painted to merge into a ground scheme, and many other schemes were devised to prevent aerial observation. Troops were moved by night for the most part, owing to the keen eyes of the air pilots and the danger of bombs, though occasionally the aviator had his chance. There is one story concerning a British pilot who, on returning from a reconnaissance flight, observed a German Staff car on the road under him; he descended and bombed and machine--gunned the car until the German General and his chauffeur abandoned it, took to their heels, and ran like rabbits. Later still, when Allied air superiority was assured, there came the phase of machine-gunning bodies of enemy troops from the air. Disregarding all antiaircraft measures, machines would sweep down and throw battalions into panic or upset the military traffic along a road, demoralising a battery or a transport train and causing as much damage through congestion of traffic as with their actual machine-gun fire. Aerial photography, too, became a fine art; the ordinary long focus cameras were used at the outset with automatic plate changers, but later on photographing aeroplanes had cameras of wide angle lens type built into the fuselage. These were very simply operated, one lever registering the exposure and changing the plate. In many cases, aerial photographs gave information which the human eye had missed, and it is noteworthy that photographs of ground showed when troops had marched over it, while the aerial observer was quite unable to detect the marks left by their passing.

Some small mention must be made of seaplane activities, which, round the European coasts involved in the War, never ceased. The submarine campaign found in the spotting seaplane its greatest deterrent, and it is old news now how even the deeply submerged submarines were easily picked out for destruction from a height and the news wirelessed from seaplane to destroyer, while in more than one place the seaplane itself finished the task by bomb dropping. It was a seaplane that gave Admiral Beatty the news that the whole German Fleet was out before the Jutland

Battle, news which led to a change of plans that very nearly brought about the destruction of Germany's naval power. For the most part, the seaplanes of the War period were heavier than the land machines and, in the opinion of the land pilots, were slow and clumsy things to fly. This was inevitable, for their work demanded more solid building and greater reliability. To put the matter into Hibernian phrase, a forced landing at sea is a much more serious matter than on the ground. Thus there was need for greater engine power, bigger wingspread to support the floats, and fuel tanks of greater capacity. The flying boats of the later War period carried considerable crews, were heavily armed, capable of withstanding very heavy weather, and carried good loads of bombs on long cruises. Their work was not all essentially seaplane work, for the R.N.A.S. was as well known as hated over the German airship sheds in Belgium and along the Flanders coast. As regards other theatres of War, they rendered valuable service from the Dardanelles to the Rufiji River, at this latter place forming a principal factor in the destruction of the cruiser Konigsberg. Their spotting work at the Dardanelles for the battleships was responsible for direct hits from 15 in. guns on invisible targets at ranges of over 12,000 yards. Seaplane pilots were bombing specialists, including among their targets army headquarters, ammunition dumps, railway stations, submarines and their bases, docks, shipping in German harbours, and the German Fleet at Wilhelmshaven. Dunkirk, a British seaplane base, was a sharp thorn in the German side.

Turning from consideration of the various services to the exploits of the men composing them, it is difficult to particularise. A certain inevitable prejudice even at this length of time leads one to discount the valour of pilots in the German Air Service, but the names of Boelcke, von Richthofen, and Immelmann recur as proof of the courage that was not wanting in the enemy ranks, while, however much we may decry the Gotha raids over the English coast and on London, there is no doubt that the men who undertook these raids were not deficient in the form of bravery that is of more value than the unthinking valour of a minute which, observed from the right quarter, wins a military decoration.

Yet the fact that the Allied airmen kept the air at all in the early days proved on which side personal superiority lay, for they were outnumbered, out-manoeuvred, and faced by better material than any that they themselves possessed; yet they won their fights or died. The stories of their deeds are endless; Bishop, flying alone and meeting seven German machines and crashing four; the battle of May 5th, 1915, when five heroes fought and conquered twenty-seven German machines, ranging in altitude between 12,000 and 3,000 feet, and continuing the extraordinary struggle from five until six in the evening. Captain Aizlewood, attacking five enemy machines with such reckless speed that he rammed one and still reached his aerodrome safely--these are items in a long list of feats of which the character can only be realised when it is fully comprehended that the British Air Service accounted for some 8,000 enemy machines in the course of the War. Among the French there was Captain Guynemer, who at the time of his death had brought down fifty-four enemy machines, in addition to many others of which the destruction could not be officially confirmed. There was Fonck, who brought down six machines in one day, four of them within two minutes.

There are incredible stories, true as incredible, of shattered men carrying on with their work in absolute disregard of physical injury. Major Brabazon Rees, V.C., engaged a big German battle-plane in September of 1915 and, single-handed, forced his enemy out of action. Later in his career, with a serious wound in the thigh from which blood was pouring, he kept up a fight with an enemy formation until he had not a round of ammunition left, and then returned to his aerodrome to get his wound dressed. Lieutenants Otley and Dunning, flying in the Balkans, engaged a couple of enemy machines and drove them off, but not until their petrol tank had got a hole in it and Dunning was dangerously wounded in the leg. Otley improvised a tourniquet, passed it to Dunning, and, when the latter had bandaged himself, changed from the observer's to the pilot's seat, plugged the bullet hole in the tank with his thumb and steered the machine home.

These are incidents; the full list has not been, and can never be recorded, but it goes to show that in the pilot of the War period there came to being a new type of humanity, a product of evolution which fitted a certain need. Of such was Captain West, who, engaging hostile troops, was attacked by seven machines. Early in the engagement, one of his legs was partially severed by an explosive bullet and fell powerless into the controls, rendering the machine for the time unmanageable. Lifting his disabled leg, he regained control of the machine, and although wounded in the other leg, he manoeuvred his machine so skilfully that his observer was able to get several good bursts into the enemy machines, driving them away. Then, desperately wounded as he was, Captain West brought the machine over to his own lines and landed safely. He fainted from loss of blood and exhaustion, but on regaining consciousness, insisted on writing his report. Equal to this was the exploit of Captain Barker, who, in aerial combat, was wounded in the right and left thigh and had his left arm shattered, subsequently bringing down an enemy machine in flames, and then breaking through another hostile formation and reaching the British lines.

In recalling such exploits as these, one is tempted on and on, for it seems that the pilots rivalled each other in their devotion to duty, this not confined to British aviators, but common practically to all services. Sufficient instances have been given to show the nature of the work and the character of the men who did it.

The rapid growth of aerial effort rendered it necessary in January of 1915 to organise the Royal Flying Corps into separate wings, and in October of the same year it was constituted in Brigades. In 1916 the Air Board was formed, mainly with the object of co-ordinating effort and ensuring both to the R.N.A.S. and to the R.F.C. adequate supplies of material as far as construction admitted. Under the presidency of Lord Cowdray, the Air Board brought about certain reforms early in 1917, and in November of that year a separate Air Ministry was constituted, separating the Air Force from both Navy and Army, and rendering it an independent force. On April 1st, 1918, the Royal Air Force came into existence, and unkind critics in the Royal Flying Corps remarked on the appropriateness of the date. At the end of the War, the personnel of the Royal Air Force amounted to 27,906 officers, and 263,842 other ranks. Contrast of these figures with the number of officers and men who took the field in 1914 is indicative of the magnitude of British aerial effort in the War period.

XX. THE WAR PERIOD--II

There was when War broke out no realisation on the part of the British Government of the need for encouraging the enterprise of private builders, who carried out their work entirely at their-own cost. The importance of a supply of British-built engines was realised before the War, it is true, and a competition was held in which a prize of L5,000 was offered for the best British engine, but this awakening was so late that the R.F.C. took the field without a single British power plant. Although Germany woke up equally late to the need for home produced aeroplane engines, the experience gained in building engines for dirigibles sufficed for the production of aeroplane power plants. The Mercedes filled all requirements together with the Benz and the Maybach. There was a 225 horsepower Benz which was very popular, as were the 100 horse-power and 170 horse-power Mercedes, the last mentioned fitted to the Aviatik biplane of 1917. The Uberursel was a copy of the Gnome and supplied the need for rotary engines.

In Great Britain there were a number of aeroplane constructing firms that had managed to emerge from the lean years 1912-1913 with sufficient manufacturing plant to give a hand in making up the leeway of construction when War broke out. Gradually the motor-car firms came in, turning their body-building departments to plane and fuselage construction, which enabled them to turn out the complete planes engined and ready for the field. The coach-building trade soon joined in and came in handy as propeller makers; big upholstering and furniture firms and scores of concerns that had never dreamed of engaging in aeroplane construction were busy on supplying the R.F.C. By 1915 hundreds of different firms were building aeroplanes and parts; by 1917 the number had increased to over 1,000, and a capital of over a million pounds for a firm that at the outbreak of War had employed a score or so of hands was by no means uncommon. Women and girls came into the work, more especially in plane construction and covering and doping, though they took their place in the engine shops and proved successful at acetylene welding and work at the lathes. It was some time before Britain was able to provide its own magnetos, for this key industry had been left in the hands of the Germans up to the outbreak of War, and the 'Bosch' was admittedly supreme--even now it has never been beaten, and can only be equalled, being as near perfection as is possible for a magneto.

One of the great inventions of the War was the synchronisation of engine-timing and machine gun, which rendered it possible to fire through the blades of a propeller without damaging them, though the growing efficiency of the aeroplane as a whole and of its armament is a thing to marvel at on looking back and considering what was actually accomplished. As the efficiency of the aeroplane increased, so anti-aircraft guns and range-finding were improved. Before the War an aeroplane travelling at full speed was reckoned perfectly safe at 4,000 feet, but, by the first month of 1915, the safe height had gone up to 9,000 feet, 7,000 feet being the limit of rifle and machine gun bullet trajectory; the heavier guns were not sufficiently mobile to tackle aircraft. At that time, it was reckoned that effective aerial photography ceased at 6,000 feet, while bomb-dropping from 7,000-8,000 feet was reckoned uncertain except in the case of a very large target. The improvement in anti-aircraft devices went on, and by May of 1916, an aeroplane was not safe under 15,000 feet, while anti-aircraft shells had fuses capable of being set to over 20,000 feet, and bombing from 15,000 and 16,000 feet was common. It was not till later that Allied pilots demonstrated the safety that lies in flying very near the ground, this owing to the fact that, when flying swiftly at a very low altitude, the machine is out of sight almost before it can be aimed at.

The Battle of the Somme and the clearing of the air preliminary to that operation brought the fighting aeroplane pure and simple with them. Formations of fighting planes preceded reconnaissance craft in order to clear German machines and observation balloons out of the sky and to watch and keep down any further enemy formations that might attempt to interfere with Allied observation work. The German reply to this consisted in the formation of the Flying Circus, of which Captain Baron von Richthofen's was a good example. Each circus consisted of a large formation of speedy machines, built specially for fighting and manned by the best of the German pilots. These were sent to attack at any point along the line where the Allies had got a decided superiority.

The trick flying of pre-war days soon became an everyday matter; Pegoud astonished the aviation world before the War by first looping the loop, but, before three years of hostilities had elapsed, looping was part of the training of practically every pilot, while the spinning nose dive, originally considered fatal, was mastered, and the tail slide, which consisted of a machine rising nose upward in the air and falling back on its tail, became one of the easiest 'stunts' in the pilot's repertoire. Inherent stability was gradually improved, and, from 1916 onward, practically every pilot could carry on with his machine-gun or camera and trust to his machine to fly itself until he was free to attend to it. There was more than one story of a machine coming safely to earth and making good landing on its own account with the pilot dead in his cock-pit.

Toward the end of the War, the Independent Air Force was formed as a branch of the R.A.F. with a view to bombing German bases and devoting its attention exclusively to work behind the enemy lines. Bombing operations were undertaken by the R.N.A.S. as early as 1914-1915 against Cuxhaven, Dusseldorf, and Friedrichshavn, but the supply of material was not sufficient to render these raids continuous. A separate Brigade, the 8th, was formed in 1917 to harass the German chemical and iron industries, the base being in the Nancy area, and this policy was found so fruitful that the Independent Force was constituted on the 8th June, 1918. The value of the work accomplished by this force is demonstrated by the fact that the German High Command recalled twenty fighting squadrons from the Western front to counter its activities, and, in addition, took troops away from the fighting line in large numbers for manning anti-aircraft batteries and searchlights. The German press of the last year of the War is eloquent of the damage done in manufacturing areas by the Independent Force, which, had hostilities continued a little longer, would have included Berlin in its activities.

Formation flying was first developed by the Germans, who made use of it in the daylight raids against England in 1917. Its value was very soon realised, and the V formation of wild geese was adopted, the leader taking the point of the V and his squadron following on either side at different heights. The air currents set up by the leading machines were thus avoided by those in the rear, while each pilot had a good view of the leader's bombs, and were able to correct their own aim by the bursts, while the different heights at which they flew rendered anti-aircraft gun practice less effective. Further, machines were able to afford mutual protection to each other and any attacker would be met by machine-gun fire from three or four machines firing on him from different angles and heights. In the later formations single-seater fighters flew above the bombers for the purpose of driving off hostile craft. Formation flying was not fully developed when the end of the War brought stagnation in place of the rapid advance in the strategy and tactics of military air work.

XXI. RECONSTRUCTION

The end of the War brought a pause in which the multitude of aircraft constructors found themselves faced with the possible complete stagnation of the industry, since military activities no longer demanded their services and the prospects of commercial flying were virtually nil. That great factor in commercial success, cost of plant and upkeep, had received no consideration whatever in the War period, for armies do not count cost. The types of machines that had evolved from the War were very fast, very efficient, and very expensive, although the bombers showed promise of adaptation to commercial needs, and, so far as other machines were concerned, America had already proved the possibilities of mail-carrying by maintaining a mail service even during the War period.

A civil aviation department of the Air Ministry was formed in February of 1919 with a Controller General of Civil Aviation at the head. This was organised into four branches, one dealing with the survey and preparation of air routes for the British Empire, one organising meteorological and wireless telegraphy services, one dealing with the licensing of aerodromes, machines for passenger or goods carrying and civilian pilots, and one dealing with publicity and transmission of information generally. A special Act of Parliament 264 entitled 'The Air Navigation Acts, 1911-1919,' was passed on February 27th, and commercial flying was officially permitted from May 1st, 1919.

Meanwhile the great event of 1919, the crossing of the Atlantic by air, was gradually ripening to performance. In addition to the rigid airship, R.34, eight machines entered for this flight, these being a Short seaplane, Handley-Page, Martinsyde, Vickers-Vimy, and Sopwith aeroplanes, and three American flying boats, N.C.1, N.C.3, and N.C.4. The Short seaplane was the only one of the eight which proposed to make the journey westward; in flying from England to Ireland, before starting on the long trip to Newfoundland, it fell into the sea off the coast of Anglesey, and so far as it was concerned the attempt was abandoned.

The first machines to start from the Western end were the three American seaplanes, which on the morning of May 6th left Trepassy, Newfoundland, on the

1,380 mile stage to Horta in the Azores. N.C.1 and N.C.3 gave up the attempt very early, but N.C.4, piloted by Lieut.-Commander Read, U.S.N., made Horta on May 17th and made a three days' halt. On the 20th the second stage of the journey to Ponta Delgada, a further 190 miles, was completed and a second halt of a week was made. On the 27th, the machine left for Lisbon, 900 miles distant, and completed the journey in a day. On the 30th a further stage of 340 miles took N.C.4 on to Ferrol, and the next day the last stage of 420 miles to Plymouth was accomplished.

Meanwhile, H. G. Hawker, pilot of the Sopwith biplane, together with Commander Mackenzie Grieve, R.N., his navigator, found the weather sufficiently auspicious to set out at 6.48 p.m. On Sunday, May 18th, in the hope of completing the trip by the direct route before N.C.4 could reach Plymouth. They set out from Mount Pearl aerodrome, St John's, Newfoundland, and vanished into space, being given up as lost, as Hamel was lost immediately before the War in attempting to fly the North Sea. There was a week of dead silence regarding their fate, but on the following Sunday morning there was world-wide relief at the news that the plucky attempt had not ended in disaster, but both aviators had been picked up by the steamer Mary at 9.30 a.m. on the morning of the 19th, while still about 750 miles short of the conclusion of their journey. Engine failure brought them down, and they planed down to the sea close to the Mary to be picked up; as the vessel was not fitted with wireless, the news of their rescue could not be communicated until land was reached. An equivalent of half the L10,000 prize offered by the Daily Mail for the non-stop flight was presented by the paper in recognition of the very gallant attempt, and the King conferred the Air Force Cross on both pilot and navigator.

Raynham, pilot of the Martinsyde competing machine, had the bad luck to crash his craft twice in attempting to start before he got outside the boundary of the aerodrome. The Handley-Page machine was withdrawn from the competition, and, attempting to fly to America, was crashed on the way.

The first non-stop crossing was made on June 14th-15th in 16 hours 27 minutes, the speed being just over 117 miles per hour. The machine was a Vickers-Vimy bomber, engined with two Rolls-Royce Eagle VIII's, piloted by Captain John Alcock, D.S.C., with Lieut. Arthur Whitten-Brown as navigator. The journey was reported to be very rough, so much so at times that Captain Alcock stated that they were flying upside down, and for the greater part of the time they were out of sight of the sea. Both pilot and navigator had the honour of knighthood conferred on them at the conclusion of the journey.

Meanwhile, commercial flying opened on May 8th (the official date was May 1st) with a joy-ride service from Hounslow of Avro training machines. The enterprise caught on remarkably, and the company extended their activities to coastal resorts for the holiday season--at Blackpool alone they took up 10,000 passengers before the service was two months old. Hendon, beginning passenger flights on the same date, went in for exhibition and passenger flying, and on June 21st the aerial Derby was won by Captain Gathergood on an Airco 4R machine with a Napier 450 horse-power 'Lion' engine; incidentally the speed of 129.3 miles per hour was officially recognised as constituting the world's record for speed within a closed circuit. On July 17th a Fiat B.R. biplane with a 700 horse-power engine landed at Kenley aerodrome after having made a non-stop flight of 1,100 miles. The maximum speed of this machine was 160 miles per hour, and it was claimed to be the fastest machine in existence. On August 25th a daily service between London and Paris was inaugurated by the Aircraft Manufacturing Company, Limited, who ran a machine each way each day, starting at 12.30 and due to arrive at 2.45 p.m. The Handley-Page Company began a similar service in September of 1919, but ran it on alternate days with machines capable of accommodating ten passengers. The single fare in each case was fixed at 15 guineas and the parcel rate at 7s. 6d. per pound.

Meanwhile, in Germany, a number of passenger services had been in operation from the early part of the year; the Berlin-Weimar service was established on February 5th and Berlin-Hamburg on March 1st, both for mail and passenger carrying. Berlin-Breslau was soon added, but the first route opened remained most popular, 538 flights being made between its opening and the end of April, while for March and April combined, the Hamburg-Berlin route recorded only 262 flights. All three routes were operated by a combine of German aeronautical firms entitled the Deutsch Luft Rederie. The single fare between Hamburg and Berlin was 450 marks, between Berlin and Breslau 500 marks, and between Berlin and Weimar 450 marks. Luggage was carried free of charge, but varied according to the weight of the passenger, since the combined weight of both passenger and luggage was not allowed to exceed a certain limit.

In America commercial flying had begun in May of 1918 with the mail service between Washington, Philadelphia, and New York, which proved that mail carrying is a commercial possibility, and also demonstrated the remarkable reliability of the modern aeroplane by making 102 complete flights out of a possible total of 104 in November, 1918, at a cost of 0.777 of a dollar per mile. By March of 1919 the cost per mile had gone up to 1.28 dollars; the first annual report issued at the end of May showed an efficiency of 95.6 per cent and the original six aeroplanes and engines with which the service began were still in regular use.

In June of 1919 an American commercial firm chartered an aeroplane for emergency service owing to a New York harbour strike and found it so useful that they made it a regular service. The Travellers Company inaugurated a passenger flying boat service between New York and Atlantic City on July 25th, the fare, inclusive of 35 lbs. of luggage, being fixed at L25 each way.

Five flights on the American continent up to the end of 1919 are worthy of note. On December 13th, 1918, Lieut. D. Godoy of the Chilian army left Santiago, Chili, crossed the Andes at a height of 19,700 feet and landed at Mendoza, the capital of the wine-growing province of Argentina. On April 19th, 1919, Captain E. F. White made the first non-stop flight between New York and Chicago in 6 hours 50 minutes on a D.H.4 machine driven by a twelve-cylinder Liberty engine. Early in August Major Schroeder, piloting a French Lepere machine flying at a height of 18,400 feet, reached a speed of 137 miles per hour with a Liberty motor fitted with a super-charger. Toward the end of August, Rex Marshall, on a Thomas-Morse biplane, starting from a height of 17,000 feet, made a glide of 35 miles with his engine cut off, restarting it when at a height of 600 feet above the ground. About a month later R. Rohlfe, piloting a Curtiss triplane, broke the height record by reaching 34,610 feet.

XXII. 1919-20

Into the later months of 1919 comes the flight by Captain Ross-Smith from England to Australia and the attempt to make the Cape to Cairo voyage by air. The Australian Government had offered a prize of L10,000 for the first flight from England to Australia in a British machine, the flight to be accomplished in 720 consecutive hours. Ross-Smith, with his brother, Lieut. Keith Macpherson Smith, and two mechanics, left Hounslow in a Vickers-Vimy bomber with Rolls-Royce engine on November 12th and arrived at Port Darwin, North Australia, on the 10th December, having completed the flight in 27 days 20 hours 20 minutes, thus having 51 hours 40 minutes to spare out of the 720 allotted hours.

Early in 1920 came a series of attempts at completing the journey by air between Cairo and the Cape. Out of four competitors Colonel Van Ryneveld came nearest to making the journey successfully, leaving England on a standard Vickers-Vimy bomber with Rolls-Royce engines, identical in design with the machine used by Captain Ross-Smith on the England to Australia flight. A second Vickers-Vimy was financed by the Times newspaper and a third flight was undertaken with a Handley-Page machine under the auspices of the Daily Telegraph. The Air Ministry had already prepared the route by means of three survey parties which cleared the aerodromes and landing grounds, dividing their journey into stages of 200 miles or less. Not one of the competitors completed the course, but in both this and Ross-Smith's flight valuable data was gained in respect of reliability of machines and engines, together with a mass of meteorological information.

The Handley-Page Company announced in the early months of 1920 that they had perfected a new design of wing which brought about a twenty to forty per cent improvement in lift rate in the year. When the nature of the design was made public, it was seen to consist of a division of the wing into small sections, each with its separate lift. A few days later, Fokker, the Dutch inventor, announced the construction of a machine in which all external bracing wires are obviated, the wings being of a very deep section and self-supporting. The value of these two inventions remains to be seen so far as commercial flying is concerned.

The value of air work in war, especially so far as the Colonial campaigns in which British troops are constantly being engaged is in question, was very thoroughly demonstrated in a report issued early in 1920 with reference to the successful termination of the Somaliland campaign through the intervention of the Royal Air Force, which between January 21st and the 31st practically destroyed the Dervish force under the Mullah, which had been a thorn in the side of Britain since 1907. Bombs and machine-guns did the work, destroying fortifications and bringing about the surrender of all the Mullah's following, with the exception of about seventy who made their escape.

Certain records both in construction and performance had characterised the post-war years, though as design advances and comes nearer to perfection, it is obvious that records must get fewer and farther between. The record aeroplane as regards size at the time of its construction was the Tarrant triplane, which made its first--and last--flight on May 28th, 1919. The total loaded weight was 30 tons, and the machine was fitted with six 400 horse-power engines; almost immediately after the trial flight began, the machine pitched forward on its nose and was wrecked, causing fatal injuries to Captains Dunn and Rawlings, who were aboard the machine. A second accident of similar character was that which befell the giant seaplane known as the Felixstowe Fury, in a trial flight. This latter machine was intended to be flown to Australia, but was crashed over the water.

On May 4th, 1920, a British record for flight duration and useful load was established by a commercial type Handley-Page biplane, which, carrying a load of 3,690 lbs., rose to a height of 13,999 feet and remained in the air for 1 hour 20 minutes. On May 27th the French pilot, Fronval, flying at Villacoublay in a Morane-Saulnier type of biplane with Le Rhone motor, put up an extraordinary type of record by looping the loop 962 times in 3 hours 52 minutes 10 seconds. Another record of the year of similar nature was that of two French fliers, Boussotrot and Bernard, who achieved a continuous flight of 24 hours 19 minutes 7 seconds, beating the pre-war record of 21 hours 48 3/4 seconds set up by the German pilot, Landemann. Both these records are likely to stand, being in the nature of freaks, which demonstrate little beyond the reliability of the machine and the capacity for endurance on the part of its pilots.

Meanwhile, on February 14th, Lieuts. Masiero and Ferrarin left Rome on S.V.A. Ansaldo V. machines fitted with 220 horse-power S.V.A. motors. On May 30th they arrived at Tokio, having flown by way of Bagdad, Karachi, Canton, Pekin, and Osaka. Several other competitors started, two of whom were shot down by Arabs in Mesopotamia.

Considered in a general way, the first two years after the termination of the Great European War form a period of transition in which the commercial type of aeroplane was gradually evolved from the fighting machine which was perfected in the four preceding years. There was about this period no sense of finality, but it was as experimental, in its own way, as were the years of progressing design which preceded the war period. Such commercial schemes as were inaugurated call for no more note than has been given here; they have been experimental, and, with the possible exception of the United States Government mail service, have not been planned and executed on a sufficiently large scale to furnish reliable data on which to forecast the prospects of commercial aviation. And there is a school rapidly growing up which asserts that the day of aeroplanes is nearly over. The construction of the giant airships of to-day and the successful return flight of R34 across the Atlantic seem to point to the eventual triumph, in spite of its disadvantages, of the dirigible airship.

This is a hard saying for such of the aeroplane industry as survived the War period and consolidated itself, and it is but the saying of a section which bases its belief on the fact that, as was noted in the very early years of the century, the aeroplane is primarily a war machine. Moreover, the experience of the War period tended to discredit the dirigible, since, before the introduction of helium gas, the inflammability of its buoyant factor placed it at an immense disadvantage beside the machine dependent on the atmosphere itself for its lift.

As life runs to-day, it is a long time since Kipling wrote his story of the airways of a future world and thrust out a prophecy that the bulk of the world's air traffic would be carried by gas-bag vessels. If the school which inclines to belief in the dirigible is right in its belief, as it well may be, then the foresight was uncannily correct, not only in the matter of the main assumption, but in the detail with which the writer embroidered it.

On the constructional side, the history of the aeroplane is still so much in the making that any attempt at a critical history would be unwise, and it is possible only to record fact, leaving it to the future for judgment to be passed. But, in a general way, criticism may be advanced with regard to the place that aeronautics takes in civilisation. In the past hundred years, the world has made miraculously rapid strides materially, but moral development has not kept abreast. Conception of the responsibilities of humanity remains virtually in a position of a hundred years ago; given a higher conception of life and its responsibilities, the aeroplane becomes the crowning achievement of that long series which James Watt inaugurated, the last step in intercommunication, the chain with which all nations are bound in a growing prosperity, surely based on moral wellbeing. Without such conception of the duties as well as the rights of life, this last achievement of science may yet prove the weapon that shall end civilisation as men know it to-day, and bring this ultra-material age to a phase of ruin on which saner people can build a world more reasonable and less given to groping after purely material advancement.

PART II. 1903-1920: PROGRESS IN DESIGN
By Lieut.-Col. W. Lockwood Marsh

I. THE BEGINNINGS

Although the first actual flight of an aeroplane was made by the Wrights on December 17th 1903, it is necessary, in considering the progress of design between that period and the present day, to go back to the earlier days of their experiments with 'gliders,' which show the alterations in design made by them in their step-bystep progress to a flying machine proper, and give a clear idea of the stage at which they had arrived in the art of aeroplane design at the time of their first flights.

They started by carefully surveying the work of previous experimenters, such as Lilienthal and Chanute, and from the lesson of some of the failures of these pioneers evolved certain new principles which were embodied in their first glider, built in 1900. In the first place, instead of relying upon the shifting of the operator's body to obtain balance, which had proved too slow to be reliable, they fitted in front of the main supporting surfaces what we now call an 'elevator,' which could be flexed, to control the longitudinal balance, from where the operator lay prone upon the main supporting surfaces. The second main innovation which they incorporated in this first glider, and the principle of which is still used in every aeroplane in existence, was the attainment of lateral balance by warping the extremities of the main planes. The effect of warping or pulling down the extremity of the wing on one side was to increase its lift and so cause that side to rise. In the first two gliders this control was also used for steering to right and left. Both these methods of control were novel for other than model work, as previous experimenters, such as Lilienthal and Pilcher, had relied entirely upon moving the legs or shifting the position of the body to control the longitudinal and lateral motions of their gliders. For the main supporting surfaces of the glider the biplane system of Chanute's gliders was adopted with certain modifications, while the curve of the wings was founded upon the calculations of Lilienthal as to wind pressure and consequent lift of the plane.

This first glider was tested on the Kill Devil Hill sand-hills in North Carolina in the summer of 1900 and proved at any rate the correctness of the principles of the front elevator and warping wings, though its designers were puzzled by the fact that the lift was less than they expected; whilst the 'drag'(as we call it), or resistance, was also considerably lower than their predictions. The 1901 machine was, in consequence, nearly doubled in area--the lifting surface being increased from 165 to 308 square feet--the first trial taking place on July 27th, 1901, again at Kill Devil Hill. It immediately appeared that something was wrong, as the machine dived straight to the ground, and it was only after the operator's position had been moved nearly a foot back from what had been calculated as the correct position that the machine would glide--and even then the elevator had to be used far more strongly than in the previous year's glider. After a good deal of thought the apparent solution of the trouble was finally found.

This consisted in the fact that with curved surfaces, while at large angles the centre of pressure moves forward as the angle decreases, when a certain limit of angle is reached it travels suddenly backwards and causes the machine to dive. The Wrights had known of this tendency from Lilienthal's researches, but had imagined that the phenomenon would disappear if they used a fairly lightly cambered--or curved--surface with a very abrupt curve at the front. Having discovered what appeared to be the cause they surmounted the difficulty by 'trussing down' the camber of the wings, with the result that they at once got back to the old conditions of the previous year and could control the machine readily with small movements of the elevator, even being able to follow undulations in the ground. They still found, however, that the lift was not as great as it should have been; while the drag remained, as in the previous glider, surprisingly small. This threw doubt on previous figures as to wind resistance and pressure on curved surfaces; but at the same time confirmed (and this was a most important result) Lilienthal's previously questioned theory that at small angles the pressure on a curved surface instead of being normal, or at right angles to, the chord is in fact inclined in front of the perpendicular. The result of this is that the pressure actually tends to draw the machine forward into the wind--hence the small amount of drag, which had puzzled Wilbur and Orville Wright.

Another lesson which was learnt from these first two years of experiment, was that where, as in a biplane, two surfaces are superposed one above the other, each of them has somewhat less lift than it would have if used alone. The experimenters were also still in doubt as to the efficiency of the warping method of controlling the lateral balance as it gave rise to certain phenomena which puzzled them, the machine turning towards the wing having the greater angle, which seemed also to touch the ground first, contrary to their expectations. Accordingly, on returning to Dayton towards the end of 1901, they set themselves to solve the various problems which had appeared and started on a lengthy series of experiments to check the previous figures as to wind resistance and lift of curved surfaces, besides setting themselves to grapple with the difficulty of lateral control. They accordingly constructed for themselves at their home in Dayton a wind tunnel 16 inches square by 6 feet long in which they measured the lift and 'drag' of more than two hundred miniature wings. In the course of these tests they for the first time produced comparative results of the lift of oblong and square surfaces, with the result that they re-discovered the importance of 'aspect ratio'--the ratio of length to breadth of planes. As a result, in the next year's glider the aspect ration of the wings was increased from the three to one of the earliest model to about six to one, which is approximately the same as that used in the machines of to-day. Further than that, they discussed the question of lateral stability, and came to the conclusion that the cause of the trouble was that the effect of warping down one wing was to increase the resistance of, and consequently slow down, that wing to such an extent that its lift was reduced sufficiently to wipe out the anticipated increase in lift resulting from the warping. From this they deduced that if the speed of the warped wing could be controlled the advantage of increasing the angle by warping could be utilised as they originally intended. They therefore decided to fit a vertical fin at the rear which, if the machine attempted to turn, would be exposed more and more to the wind and so stop the turning motion by offering increased resistance.

As a result of this laboratory research work the third Wright glider, which was taken to Kill Devil Hill in September, 1902, was far more efficient aerodynamically than either of its two predecessors, and was fitted with a fixed vertical fin at the rear in addition to the movable elevator in front. According to Mr Griffith Brewer[8], this third glider contained 305 square feet of surface; though there may possibly be a mistake here, as he states[9] the surface of the previous year's glider to have been only 290 square feet, whereas Wilbur Wright himself[10] states it to have been 308 square feet. The matter is not, perhaps, save historically, of much importance, except that the gliders are believed to have been progressively larger, and therefore if we accept Wilbur Wright's own figure of the surface of the second glider, the third must have had a greater area than that given by Mr Griffith Brewer. Unfortunately, no evidence of the Wright Brothers themselves on this point is available.

The first glide of the 1902, season was made on September 17th of that year, and the new machine at once showed itself an improvement on its predecessors, though subsequent trials showed that the difficulty of lateral balance had not been entirely overcome. It was decided, therefore, to turn the vertical fin at the rear into a rudder by making it movable. At the same time it was realised that there was a definite relation between lateral balance and directional control, and the rudder controls and wing-warping wires were accordingly connected This ended the pioneer gliding experiments of Wilbur and Orville Wright--though further glides were made in subsequent years--as the following year, 1903, saw the first power-driven machine leave the ground.

To recapitulate--in the course of these original experiments the Wrights confirmed Lilienthal's theory of the reversal of the centre of pressure on cambered surfaces at small angles of incidence: they confirmed the importance of high aspect ratio in respect to lift: they had evolved new and more accurate tables of lift and pressure on cambered surfaces: they were the first to use a movable horizontal elevator for controlling height: they were the first to adjust the wings to different

---

8   Fourth Wilbur Wright Memorial Lecture, Aeronautical Journal, Vol. XX, No. 79, page 75.

9   Ibid. page 73.

10 Ibid. pp. 91 and 102.

angles of incidence to maintain lateral balance: and they were the first to use the movable rudder and adjustable wings in combination.

They now considered that they had gone far enough to justify them in building a power-driven 'flier,' as they called their first aeroplane. They could find no suitable engine and so proceeded to build for themselves an internal combustion engine, which was designed to give 8 horse-power, but when completed actually developed about 12-15 horse-power and weighed 240 lbs. The complete machine weighed about 750 lbs. Further details of the first Wright aeroplane are difficult to obtain, and even those here given should be received with some caution. The first flight was made on December 17th 1903, and lasted 12 seconds. Others followed immediately, and the fourth lasted 59 seconds, a distance of 852 feet being covered against a 20-mile wind.

The following year they transferred operations to a field outside Dayton, Ohio (their home), and there they flew a somewhat larger and heavier machine with which on September 20th 1904, they completed the first circle in the air. In this machine for the first time the pilot had a seat; all the previous experiments having been carried out with the operator lying prone on the lower wing. This was followed next year by another still larger machine, and on it they carried out many flights. During the course of these flights they satisfied themselves as to the cause of a phenomenon which had puzzled them during the previous year and caused them to fear that they had not solved the problem of lateral control. They found that on occasions--always when on a turn--the machine began to slide down towards the ground and that no amount of warping could stop it. Finally it was found that if the nose of the machine was tilted down a recovery could be effected; from which they concluded that what actually happened was that the machine, 'owing to the increased load caused by centrifugal force,' had insufficient power to maintain itself in the air and therefore lost speed until a point was reached at which the controls became inoperative. In other words, this was the first experience of 'stalling on a turn,' which is a danger against which all embryo pilots have to guard in the early stages of their training.

The 1905 machine was, like its predecessors, a biplane with a biplane elevator in front and a double vertical rudder in rear. The span was 40 feet, the chord of the wings being 6 feet and the gap between them about the same. The total area was about 600 square feet which supported a total weight of 925 lbs.; while the motor was 12 to 15 horse-power driving two propellers on each side behind the main planes through chains and giving the machine a speed of about 30 m.p.h. one of these chains was crossed so that the propellers revolved in opposite directions to avoid the torque which it was feared would be set up if they both revolved the same way. The machine was not fitted with a wheeled undercarriage but was carried on two skids, which also acted as outriggers to carry the elevator. Consequently, a mechanical method of launching had to be evolved and the machine received initial velocity from a rail, along which it was drawn by the impetus provided by the falling of a weight from a wooden tower or 'pylon.' As a result of this the Wright aeroplane in its original form had to be taken back to its starting rail after each flight, and could not restart from the point of alighting. Perhaps, in comparison with French machines of more or less contemporary date (evolved on independent lines in ignorance of the Americans' work), the chief feature of the Wright biplane of 1905 was that it relied entirely upon the skill of the operator for its stability; whereas in France some attempt was being made, although perhaps not very successfully, to make the machine automatically stable laterally. The performance of the Wrights in carrying a loading of some 60 lbs. per horse-power is one which should not be overlooked. The wing loading was about 1 1/2 lbs. per square foot.

About the same time that the Wrights were carrying out their power-driven experiments, a band of pioneers was quite independently beginning to approach success in France. In practically every case, however, they started from a somewhat different standpoint and took as their basic idea the cellular (or box) kite. This form of kite, consisting of two superposed surfaces connected at each end by a vertical panel or curtain of fabric, had proved extremely successful for man-carrying purposes, and, therefore, it was little wonder that several minds conceived the idea of attempting to fly by fitting a series of box-kites with an engine. The first to achieve success was M. Santos-Dumont, the famous Brazilian pioneer-designer of airships, who, on November 12th, 1906, made several flights, the last of which covered a little over 700 feet. Santos-Dumont's machine consisted essentially of two box-kites, forming the main wings, one on each side of the body, in which the pilot stood, and at the front extremity of which was another movable box-kite to act as elevator and rudder. The curtains at the ends were intended to give lateral stability, which was further ensured by setting the wings slightly inclined upwards from the centre, so that when seen from the front they formed a wide V. This feature is still to be found in many aeroplanes to-day and has come to be known as the 'dihedral.' The motor was at first of 24 horse-power, for which later a 50 horse-power Antoinette engine was substituted; whilst a three-wheeled undercarriage was provided, so that the machine could start without external mechanical aid. The machine was constructed of bamboo and steel, the weight being as low as 352 lbs. The span was 40 feet, the length being 33 feet, with a total surface of main planes of 860 square feet. It will thus be seen--for comparison with the Wright machine--that the weight per horse-power (with the 50 horse-power engine) was only 7 lbs., while the wing loading was equally low at 1/2 lb. per square foot.

The main features of the Santos-Dumont machine were the box-kite form of construction, with a dihedral angle on the main planes, and the forward elevator which could be moved in any direction and therefore acted in the same way as the rudder at the rear of the Wright biplane. It had a single propeller revolving in the centre behind the wings and was fitted with an undercarriage incorporated in the machine.

The other chief French experimenters at this period were the Voisin Freres, whose first two machines--identical in form--were sold to Delagrange and H. Farman, which has sometimes caused confusion, the two purchasers being credited with the design they bought. The Voisins, like the Wrights, based their designs largely on the experimental work of Lilienthal, Langley, Chanute, and others, though they also carried out tests on the lifting properties of aerofoils in a wind tunnel of their own. Their first machines, like those of Santos-Dumont, showed the effects of experimenting with box-kites, some of which they had built for M. Ernest Archdeacon in 1904. In their case the machine, which was again a biplane, had, like both the others previously mentioned, an elevator in front--though in this case of monoplane form--and, as in the Wright, a rudder was fitted in rear of the main planes. The Voisins, however, fitted a fixed biplane horizontal 'tail'--in an effort to obtain a measure of automatic longitudinal stability--between the two surfaces of which the single rudder worked. For lateral stability they depended entirely on end curtains between the upper and lower surfaces of both the main planes and biplane tail surfaces. They, like Santos-Dumont, fitted a wheeled undercarriage, so that the machine was self-contained. The Voisin machine, then, was intended to be automatically stable in both senses; whereas the Wrights deliberately produced a machine which was entirely dependent upon the pilot's skill for its stability. The dimensions of the Voisin may be given for comparative purposes, and were as follows: Span 33 feet with a chord (width from back to front) of main planes of 6 1/2 feet, giving a total area of 430 square feet. The 50 horse-power Antoinette engine, which was enclosed in the body (or 'nacelle ') in the front of which the pilot sat, drove a propeller behind, revolving between the outriggers carrying the tail. The total weight, including Farman as pilot, is given as 1,540 lbs., so that the machine was much heavier than either of the others; the weight per horse-power being midway between the Santos-Dumont and the Wright at 31 lbs. per square foot, while the wing loading was considerably greater than either at 3 1/2 lbs. per square foot. The Voisin machine was experimented with by Farman and Delagrange from about June 1907 onwards, and was in the subsequent years developed by Farman; and right up to the commencement of the War upheld the principles of the box-kite method of construction for training purposes. The chief modification of the original design was the addition of flaps (or ailerons) at the rear extremities of the main planes to give lateral control, in a manner analogous to the wing-warping method invented by the Wrights, as a result of which the end curtains between the planes were abolished. An additional elevator was fitted at the rear of the fixed biplane tail, which eventually led to the discarding of the front elevator altogether. During the same period the Wright machine came into line with the others by the fitting of a wheeled undercarriage integral with the machine. A fixed horizontal tail was also added to the rear rudder, to which a movable elevator was later attached; and, finally, the front elevator was done away with. It will thus be seen that having started from the very different standpoints of automatic stability and complete control by the pilot, the Voisin (as developed in the Farman) and Wright machines, through gradual evolution finally resulted in aeroplanes of similar characteristics embodying a modicum of both features.

Before proceeding to the next stage of progress mention should be made of the experimental work of Captain Ferber in France. This officer carried out a large number of experiments with gliders contemporarily with the Wrights, adopting--like them--the Chanute biplane principle. He adopted the front elevator from the

Wrights, but immediately went a step farther by also fitting a fixed tail in rear, which did not become a feature of the Wright machine until some seven or eight years later. He built and appeared to have flown a machine fitted with a motor in 1905, and was commissioned to go to America by the French War Office on a secret mission to the Wrights. Unfortunately, no complete account of his experiments appears to exist, though it can be said that his work was at least as important as that of any of the other pioneers mentioned.

II. MULTIPLICITY OF IDEAS

In a review of progress such as this, it is obviously impossible, when a certain stage of development has been reached, owing to the very multiplicity of experimenters, to continue dealing in anything approaching detail with all the different types of machines; and it is proposed, therefore, from this point to deal only with tendencies, and to mention individuals merely as examples of a class of thought rather than as personalities, as it is often difficult fairly to allocate the responsibility for any particular innovation.

During 1907 and 1908 a new type of machine, in the monoplane, began to appear from the workshops of Louis Bleriot, Robert Esnault-Pelterie, and others, which was destined to give rise to long and bitter controversies on the relative advantages of the two types, into which it is not proposed to enter here; though the rumblings of the conflict are still to be heard by discerning ears. Bleriot's early monoplanes had certain new features, such as the location of the pilot, and in some cases the engine, below the wing; but in general his monoplanes, particularly the famous No. XI on which the first Channel crossing was made on July 25th, 1909, embodied the main principles of the Wright and Voisin types, except that the propeller was in front of instead of behind the supporting surfaces, and was, therefore, what is called a 'tractor' in place of the then more conventional 'pusher.' Bleriot aimed at lateral balance by having the tip of each wing pivoted, though he soon fell into line with the Wrights and adopted the warping system. The main features of the design of Esnault-Pelterie's monoplane was the inverted dihedral (or kathedral as this was called in Mr S. F. Cody's British Army Biplane of 1907) on the wings, whereby the tips were considerably lower than the roots at the body. This was designed to give automatic lateral stability, but, here again, conventional practice was soon adopted and the R.E.P. monoplanes, which became well-known in this country through their adoption in the early days by Messrs Vickers, were of the ordinary monoplane design, consisting of a tractor propeller with wire-stayed wings, the pilot being in an enclosed fuselage containing the engine in front and carrying at its rear extremity fixed horizontal and vertical surfaces combined with movable elevators and rudder. Constructionally, the R.E.P. monoplane was of extreme interest as the body was constructed of steel. The Antoinette monoplane, so ably flown by Latham, was another very famous machine of the 1909-1910 period, though its performance were frequently marred by engine failure; which was indeed the bugbear of all these early experimenters, and it is difficult to say, after this lapse of time, how far in many cases the failures which occurred, both in performances and even in the actual ability to rise from the ground, were due to defects in design or merely faults in the primitive engines available. The Antoinette aroused admiration chiefly through its graceful, birdlike lines, which have probably never been equalled; but its chief interest for our present purpose lies in the novel method of wing-staying which was employed. Contemporary monoplanes practically all had their wings stayed by wires to a post in the centre above the fuselage, and, usually, to the undercarriage below. In the Antoinette, however, a king post was introduced half-way along the wing, from which wires were carried to the ends of the wings and the body. This was intended to give increased strength and permitted of a greater wing-spread and consequently improved aspect ratio. The same system of construction was adopted in the British Martinsyde monoplanes of two or three years later.

This period also saw the production of the first triplane, which was built by A. V. Roe in England and was fitted with a J.A.P. engine of only 9 horse-power--an amazing performance which remains to this day unequalled. Mr Roe's triplane was chiefly interesting otherwise for the method of maintaining longitudinal control, which was achieved by pivoting the whole of the three main planes so that their angle of incidence could be altered. This was the direct converse of the universal practice of elevating by means of a subsidiary surface either in front or rear of the main planes.

Recollection of the various flying meetings and exhibitions which one attended during the years from 1909 to 1911, or even 1912 are chiefly notable for the fact that the first thought on seeing any new type of machine was not as to what its 'performance'--in speed, lift, or what not--would be; but speculation as to whether it would leave the ground at all when eventually tried. This is perhaps the best indication of the outstanding characteristic of that interim period between the time of the first actual flights and the later period, commencing about 1912, when ideas had become settled and it was at last becoming possible to forecast on the drawing-board the performance of the completed machine in the air. Without going into details, for which there is no space here, it is difficult to convey the correct impression of the chaotic state which existed as to even the elementary principles of aeroplane design. All the exhibitions contained large numbers--one had almost written a majority--of machines which embodied the most unusual features and which never could, and in practice never did, leave the ground. At the same time, there were few who were sufficiently hardy to say certainly that this or that innovation was wrong; and consequently dozens of inventors in every country were conducting isolated experiments on both good and bad lines. All kinds of devices, mechanical and otherwise, were claimed as the solution of the problem of stability, and there was even controversy as to whether any measure of stability was not undesirable; one school maintaining that the only safety lay in the pilot having the sole say in the attitude of the machine at any given moment, and fearing danger from the machine having any mind of its own, so to speak. There was, as in most controversies, some right on both sides, and when we come to consider the more settled period from 1912 to the outbreak of the War in 1914 we shall find how a compromise was gradually effected.

At the same time, however, though it was at the time difficult to pick out, there was very real progress being made, and, though a number of 'freak' machines fell out by the wayside, the pioneer designers of those days learnt by a process of trial and error the right principles to follow and gradually succeeded in getting their ideas crystallised.

In connection with stability mention must be made of a machine which was evolved in the utmost secrecy by Mr J. W. Dunne in a remote part of Scotland under subsidy from the War office. This type, which was constructed in both monoplane and biplane form, showed that it was in fact possible in 1910 and 1911 to design an aeroplane which could definitely be left to fly itself in the air. One of the Dunne machines was, for example flown from Farnborough to Salisbury Plain without any control other than the rudder being touched; and on another occasion it flew a complete circle with all controls locked automatically assuming the correct bank for the radius of turn. The peculiar form of wing used, the camber of which varied from the root to the tip, gave rise however, to a certain loss in efficiency, and there was also a difficulty in the pilot assuming adequate control when desired. Other machines designed to be stable--such as the German Etrich and the British Weiss gliders and Handley-Page monoplanes--were based on the analogy of a wing attached to a certain seed found in Nature (the 'Zanonia' leaf), on the righting effect of back-sloped wings combined with upturned (or 'negative') tips. Generally speaking, however, the machines of the 1909-1912 period relied for what automatic stability they had on the principle of the dihedral angle, or flat V, both longitudinally and laterally. Longitudinally this was obtained by setting the tail at a slightly smaller angle than the main planes.

The question of reducing the resistance by adopting 'stream-line' forms, along which the air could flow uninterruptedly without the formation of eddies, was not at first properly realised, though credit should be given to Edouard Nieuport, who in 1909 produced a monoplane with a very large body which almost completely enclosed the pilot and made the machine very fast, for those days, with low horse-power. On one of these machines C. T. Weyman won the Gordon-Bennett Cup for America in 1911 and another put up a fine performance in the same race with only a 30 horse-power engine. The subject, was however, early taken up by the British Advisory Committee for Aeronautics, which was established by the Government in 1909, and designers began to realise the importance of streamline struts and fuselages towards the end of this transition period. These efforts were at first not always successful and showed at times a lack of understanding of the problems involved, but there was a very marked improvement during the year 1912. At the Paris Aero Salon held early in that year there was a notable variety of ideas on the subject; whereas by the time of the one held in October designs had considerably settled down, more than one exhibitor showing what were called 'monocoque' fuselages completely circular in shape and having very low resistance, while the same show saw the introduction of rotating cowls over the propeller bosses, or 'spinners,' as they came to be called during the War. A particularly fine example of stream-lining was to be found in the Deperdussin monoplane on which Vedrines won back the Gordon-Bennett Aviation Cup from America at a speed of 105.5 m.p.h.--a considerable improvement on the 78 m.p.h. of the preceding year, which was by no means accounted for by the mere increase in engine power from 100 horse-power to 140 horse-power. This machine was the first in which the refinement of 'stream-lining' the pilot's head, which became a feature of subsequent racing machines, was introduced. This consisted of a circular padded excresence above the cockpit immediately behind the pilot's head, which gradually tapered off into the top surface of the fuselage. The object was to give the air an uninterrupted flow instead of allowing it to be broken up into eddies behind the head of the pilot, and it also provided a support against the enormous wind-pressure encountered. This true stream-line form of fuselage owed its introduction to the Paulhan-Tatin 'Torpille' monoplane of the Paris Salon of early 1917. Altogether the end of the year 1912 began to see the disappearance of 'freak' machines with all sorts of original ideas for the increase of stability and performance. Designs had by then gradually become to a considerable extent standardised, and it had become unusual to find a machine built which would fail to fly. The Gnome engine held the field owing to its advantages, as the first of the rotary type, in lightness and ease of fitting into the nose of a fuselage. The majority of machines were tractors (propeller in front) although a preference, which died down subsequently, was still shown for the monoplane over the biplane. This year also saw a great increase in the number of seaplanes, although the 'flying boat' type had only appeared at intervals and the vast majority were of the ordinary aeroplane type fitted with floats in place of the land undercarriage; which type was at that time commonly called 'hydro-aeroplane.' The usual horse power was 50--that of the smallest Gnome engine--although engines of 100 to 140 horse-power were also fitted occasionally. The average weight per horse-power varied from 18 to 25 lbs., while the wing-loading was usually in the neighbourhood of 5 to 6 lbs. per square foot. The average speed ranged from 65-75 miles per hour.

III. PROGRESS ON STANDARDISED LINES

In the last section an attempt has been made to show how, during what was from the design standpoint perhaps the most critical period, order gradually became evident out of chaos, ill-considered ideas dropped out through failure to make good, and, though there was still plenty of room for improvement in details, the bulk of the aeroplanes showed a general similarity in form and conception. There was still a great deal to be learnt in finding the best form of wing section, and performances were still low; but it had become definitely possible to say that flying had emerged from the chrysalis stage and had become a science. The period which now began was one of scientific development and improvement--in performance, manoeuvrability, and general airworthiness and stability.

The British Military Aeroplane Competition held in the summer of 1912 had done much to show the requirements in design by giving possibly the first opportunity for a definite comparison of the performance of different machines as measured by impartial observers on standard lines--albeit the methods of measuring were crude. These showed that a high speed--for those days--of 75 miles an hour or so was attended by disadvantages in the form of an equally fast low speed, of 50 miles per hour or more, and generally may be said to have given designers an idea what to aim for and in what direction improvements were required. In fact, the most noticeable point perhaps of the machines of this time was the marked manner in which a machine that was good in one respect would be found to be wanting in others. It had not yet been possible to combine several desirable attributes in one machine. The nearest approach to this was perhaps to be found in the much discussed Government B.E.2 machine, which was produced from the Royal Aircraft Factory at Farnborough, in the summer of 1912. Though considerably criticized from many points of view it was perhaps the nearest approach to a machine of all-round efficiency that had up to that date appeared. The climbing rate, which subsequently proved so important for military purposes, was still low, seldom, if ever, exceeding 400 feet per minute; while gliding angles (ratio of descent to forward travel over the ground with engine stopped) little exceeded 1 in 8.

The year 1912 and 1913 saw the subsequently all-conquering tractor biplane begin to come into its own. This type, which probably originated in England, and at any rate attained to its greatest excellence prior to the War from the drawing offices of the Avro Bristol and Sopwith firms, dealt a blow at the monoplane from which the latter never recovered.

The two-seater tractor biplane produced by Sopwith and piloted by H. G. Hawker, showed that it was possible to produce a biplane with at least equal speed to the best monoplanes, whilst having the advantage of greater strength and lower landing speeds. The Sopwith machine had a top speed of over 80 miles an hour while landing as slowly as little more than 30 miles an hour; and also proved that it was possible to carry 3 passengers with fuel for 4 hours' flight with a motive power of only 80 horse-power. This increase in efficiency was due to careful attention to detail in every part, improved wing sections, clean fuselage-lines, and simplified undercarriages. At the same time, in the early part of 1913 a tendency manifested itself towards the four-wheeled undercarriage, a pair of smaller wheels being added in front of the main wheels to prevent overturning while running on the ground; and several designs of oleo-pneumatic and steel-spring undercarriages were produced in place of the rubber shock-absorber type which had up till then been almost universal.

These two statements as to undercarriage designs may appear to be contradictory, but in reality they do not conflict as they both showed a greater attention to the importance of good springing, combined with a desire to avoid complication and a mass of struts and wires which increased head resistance.

The Olympia Aero Show of March, 1913, also produced a machine which, although the type was not destined to prove the best for the purpose for which it was designed, was of interest as being the first to be designed specially for war purposes. This was the Vickers 'Gun-bus,' a 'pusher' machine, with the propeller revolving behind the main planes between the outriggers carrying the tail, with a seat right in front for a gunner who was provided with a machine gun on a swivelling mount which had a free field of fire in every direction forward. The device which proved the death-blow for this type of aircraft during the war will be dealt with in the appropriate place later, but the machine should not go unrecorded.

As a result of a number of accidents to monoplanes the Government appointed a Committee at the end of 1912 to inquire into the causes of these. The report which was presented in March, 1913, exonerated the monoplane by coming to the conclusion that the accidents were not caused by conditions peculiar to monoplanes, but pointed out certain desiderata in aeroplane design generally which are worth recording. They recommended that the wings of aeroplanes should be so internally braced as to have sufficient strength in themselves not to collapse if the external bracing wires should give way. The practice, more common in monoplanes than biplanes, of carrying important bracing wires from the wings to the undercarriage was condemned owing to the liability of damage from frequent landings. They also pointed out the desirability of duplicating all main wires and their attachments, and of using stranded cable for control wires. Owing to the suspicion that one accident at least had been caused through the tearing of the fabric away from the wing, it was recommended that fabric should be more securely fastened to the ribs of the wings, and that devices for preventing the spreading of tears should be considered. In the last connection it is interesting to note that the French Deperdussin firm produced a fabric wing-covering with extra strong threads run at right-angles through the fabric at intervals in order to limit the tearing to a defined area.

In spite, however, of the whitewashing of the monoplane by the Government Committee just mentioned, considerable stir was occasioned later in the year by the decision of the War office not to order any more monoplanes; and from this time forward until the War period the British Army was provided exclusively with biplanes. Even prior to this the popularity of the monoplane had begun to wane. At the Olympia Aero Show in March, 1913, biplanes for the first time outnumbered the 'single-deckers'(as the Germans call monoplanes); which had the effect of reducing the wing-loading. In the case of the biplanes exhibited this averaged about 4 1/2 lbs. per square foot, while in the case of the monoplanes in the same exhibition the lowest was 5 1/2 lbs., and the highest over 8 1/2 lbs. per square foot of area. It may here be mentioned that it was not until the War period that the importance of loading per horse-power was recognised as the true criterion of aeroplane efficiency, far greater interest being displayed in the amount of weight borne per unit area of wing.

An idea of the state of development arrived at about this time may be gained from the fact that the Commandant of the Military Wing of the Royal Flying Corps in a lecture before the Royal Aeronautical Society read in February, 1913, asked for single-seater scout aeroplanes with a speed of 90 miles an hour and a landing speed of 45 miles an hour--a performance which even two years later would have been considered modest in the extreme. It serves to show that, although higher performances were put up by individual machines on occasion, the general development had not yet reached the stage when such performances could be obtained in machines suitable for military purposes. So far as seaplanes were concerned, up to the beginning of 1913 little attempt had been made to study the novel problems involved, and the bulk of the machines at the Monaco Meeting in April, 1913, for instance, consisted of land machines fitted with floats, in many cases of a most primitive nature, without other alterations. Most of those which succeeded in leaving the water did so through sheer pull of engine power; while practically all were incapable of getting off except in a fair sea, which enabled the pilot to jump the machine into the air across the trough between two waves. Stability problems had not yet been considered, and in only one or two cases was fin area added at the rear high up, to counterbalance the effect of the floats low down in front. Both twin and single-float machines were used, while the flying boat was only just beginning to come into being from the workshops of Sopwith in Great Britain, Borel-Denhaut in France, and Curtiss in America. In view of the approaching importance of amphibious seaplanes, mention should be made of the flying boat (or 'bat boat' as it was called, following Rudyard Kipling) which was built by Sopwith in 1913 with a wheeled landing-carriage which could be wound up above the bottom surface of the boat so as to be out of the way when alighting on water.

During 1913 the (at one time almost universal) practice originated by the Wright Brothers, of warping the wings for lateral stability, began to die out and the bulk of aeroplanes began to be fitted with flaps (or 'ailerons') instead. This was a distinct change for the better, as continually warping the wings by bending down the extremities of the rear spars was bound in time to produce 'fatigue' in that member and lead to breakage; and the practice became completely obsolete during the next two or three years.

The Gordon-Bennett race of September, 1913, was again won by a Deperdussin machine, somewhat similar to that of the previous year, but with exceedingly small wings, only 107 square feet in area. The shape of these wings was instructive as showing how what, from the general utility point of view, may be disadvantageous can, for a special purpose, be turned to account. With a span of 21 feet, the chord was 5 feet, giving the inefficient 'aspect ratio' of slightly over 4 to 1 only. The object of this was to reduce the lift, and therefore the resistance, to as low a point as possible. The total weight was 1,500 lbs., giving a wing-loading of 14 lbs. per square foot--a hitherto undreamt-of figure. The result was that the machine took an enormously long run before starting; and after touching the ground on landing ran for nearly a mile before stopping; but she beat all records by attaining a speed of 126 miles per hour. Where this performance is mainly interesting is in contrast to the machines of 1920, which with an even higher speed capacity would yet be able to land at not more than 40 or 50 miles per hour, and would be thoroughly efficient flying machines.

The Rheims Aviation Meeting, at which the Gordon-Bennett race was flown, also saw the first appearance of the Morane 'Parasol' monoplane. The Morane monoplane had been for some time an interesting machine as being the only type which had no fixed surface in rear to give automatic stability, the movable elevator being balanced through being hinged about one-third of the way back from the front edge. This made the machine difficult to fly except in the hands of experts, but it was very quick and handy on the controls and therefore useful for racing purposes. In the 'Parasol' the modification was introduced of raising the wing above the body, the pilot looking out beneath it, in order to give as good a view as possible.

Before passing to the year 1914 mention should be made of the feat performed by Nesteroff, a Russian, and Pegoud, a French pilot, who were the first to demonstrate the possibilities of flying upside-down and looping the loop. Though perhaps not coming strictly within the purview of a chapter on design (though certain alterations were made to the top wing-bracing of the machine for this purpose) this performance was of extreme importance to the development of aviation by showing the possibility of recovering, given reasonable height, from any position in the air; which led designers to consider the extra stresses to which an aeroplane might be subjected and to take steps to provide for them by increasing strength where necessary.

When the year 1914 opened a speed of 126 miles per hour had been attained and a height of 19,600 feet had been reached. The Sopwith and Avro (the forerunner of the famous training machine of the War period) were probably the two leading tractor biplanes of the world, both two-seaters with a speed variation from 40 miles per hour up to some 90 miles per hour with 80 horse-power engines. The French were still pinning their faith mainly to monoplanes, while the Germans were beginning to come into prominence with both monoplanes and biplanes of the 'Taube' type. These had wings swept backward and also upturned at the wing-tips which, though it gave a certain measure of automatic stability, rendered the machine somewhat clumsy in the air, and their performances were not on the whole as high as those of either France or Great Britain.

Early in 1914 it became known that the experimental work of Edward Busk--who was so lamentably killed during an experimental flight later in the year--following upon the researches of Bairstow and others had resulted in the production at the Royal Aircraft Factory at Farnborough of a truly automatically stable aeroplane. This was the 'R.E.' (Reconnaissance Experimental), a development of the B.E. which has already been referred to. The remarkable feature of this design was that there was no particular device to which one could point out as the cause of the stability. The stable result was attained simply by detailed design of each part of the aeroplane, with due regard to its relation to, and effect on, other parts in the air. Weights and areas were so nicely arranged that under practically any conditions the machine tended to right itself. It did not, therefore, claim to be a machine which it was impossible to upset, but one which if left to itself would tend to right itself from whatever direction a gust might come. When the principles were extended to the 'B.E. 2c' type (largely used at the outbreak of the War) the latter machine, if the engine were switched off at a height of not less than 1,000 feet above the ground, would after a few moments assume its correct gliding angle and glide down to the ground.

The Paris Aero Salon of December, 1913, had been remarkable chiefly for the large number of machines of which the chassis and bodywork had been constructed of steel-tubing; for the excess of monoplanes over biplanes; and (in the latter) predominance of 'pusher' machines (with propeller in rear of the main planes) compared with the growing British preference for 'tractors' (with air screw in front). Incidentally, the Maurice Farman, the last relic of the old type box-kite with elevator in front appeared shorn of this prefix, and became known as the 'short-horn' in contradistinction to its front-elevatored predecessor which, owing to its general reliability and easy flying capabilities, had long been affectionately called the 'mechanical cow.' The 1913 Salon also saw some lingering attempts at attaining automatic stability by pendulum and other freak devices.

Apart from the appearance of 'R.E.1,' perhaps the most notable development towards the end of 1913 was the appearance of the Sopwith 'Tabloid 'tractor biplane. This single-seater machine, evolved from the two-seater previously referred to, fitted with a Gnome engine of 80 horse-power, had the, for those days, remarkable speed of 92 miles an hour; while a still more notable feature was that it could remain in level flight at not more than 37 miles per hour. This machine is of particular importance because it was the prototype and forerunner of the successive designs of single-seater scout fighting machines which were used so extensively from 1914 to 1918. It was also probably the first machine to be capable of reaching a height of 1,000 feet within one minute. It was closely followed by the 'Bristol Bullet,' which was exhibited at the Olympia Aero Show of March, 1914. This last pre-war show was mainly remarkable for the good workmanship displayed--rather than for any distinct advance in design. In fact, there was a notable diversity in the types displayed, but in detailed design considerable improvements were to be seen, such as the general adoption of stranded steel cable in place of piano wire for the mail bracing.

IV. THE WAR PERIOD

Up to this point an attempt has been made to give some idea of the progress that was made during the eleven years that had elapsed since the days of the Wrights' first flights. Much advance had been made and aeroplanes had settled down, superficially at any rate, into more or less standardised forms in three main types--tractor monoplanes, tractor biplanes, and pusher biplanes. Through the application of the results of experiments with models in wind tunnels to full-scale machines, considerable improvements had been made in the design of wing sections, which had greatly increased the efficiency of aeroplanes by raising the amount of 'lift' obtained from the wing compared with the 'drag' (or resistance to forward motion) which the same wing would cause. In the same way the shape of bodies, interplane struts, etc., had been improved to be of better stream-line shape, for the further reduction of resistance; while the problems of stability were beginning to be tolerably well understood. Records (for what they are worth) stood at 21,000 feet as far as height was concerned, 126 miles per hour for speed, and 24 hours duration. That there was considerable room for development is, however, evidenced by a statement made by the late B. C. Hucks (the famous pilot) in the course of an address delivered before the Royal Aeronautical Society in July, 1914. 'I consider,' he said, 'that the present day standard of flying is due far more to the improvement in piloting than to the improvement in machines.... I consider those (early 1914) machines are only slight improvements on the machines of three years ago, and yet they are put through evolutions which, at that time, were not even dreamed of. I can take a good example of the way improvement in piloting has outdistanced improvement in machines--in the case of myself, my 'looping' Bleriot. Most of you know that there is very little difference between that machine and the 50 horse-power Bleriot of three years ago.' This statement was, of course, to some extent an exaggeration and was by no means agreed with by designers, but there was at the same time a germ of truth in it. There is at any rate little doubt that the theory and practice of aeroplane design made far greater strides towards becoming an exact science during the four years of War than it had done during the six or seven years preceding it.

It is impossible in the space at disposal to treat of this development even with the meagre amount of detail that has been possible while covering the 'settling down' period from 1911 to 1914, and it is proposed, therefore, to indicate the improvements by sketching briefly the more noticeable difference in various respects between the average machine of 1914 and a similar machine of 1918.

In the first place, it was soon found that it was possible to obtain greater efficiency and, in particular, higher speeds, from tractor machines than from pusher machines with the air screw behind the main planes. This was for a variety of reasons connected with the efficiency of propellers and the possibility of reducing resistance to a greater extent in tractor machines by using a 'stream-line' fuselage (or body) to connect the main planes with the tail. Full advantage of this could not be taken, however, owing to the difficulty of fixing a machine-gun in a forward direction owing to the presence of the propeller. This was finally overcome by an ingenious device (known as an 'Interrupter gear') which allowed the gun to fire only when none of the propeller blades was passing in front of the muzzle. The monoplane gradually fell into desuetude, mainly owing to the difficulty of making that type adequately strong without it becoming prohibitively heavy, and also because of its high landing speed and general lack of manoeuvrability. The triplane was also little used except in one or two instances, and, practically speaking, every machine was of the biplane tractor type.

A careful consideration of the salient features leading to maximum efficiency in aeroplanes--particularly in regard to speed and climb, which were the two most important military requirements--showed that a vital feature was the reduction in the amount of weight lifted per horse-power employed; which in 1914 averaged from 20 to 25 lbs. This was effected both by gradual increase in the power and size of the engines used and by great improvement in their detailed design (by increasing compression ratio and saving weight whenever possible); with the result that the motive power of single-seater aeroplanes rose from 80 and 100 horse-power in 1914 to an average of 200 to 300 horse-power, while the actual weight of the engine fell from 3 1/2-4 lbs. per horse-power to an average of 2 1/2 lbs. per horse-power. This meant that while a pre-war engine of 100 horse-power would weigh some 400 lbs., the 1918 engine developing three times the power would have less than double the weight. The result of this improvement was that a scout aeroplane at the time of the Armistice would have 1 horse-power for every 8 lbs. of weight lifted, compared with the 20 or 25 lbs. of its 1914 predecessors. This produced a considerable increase in the rate of climb, a good postwar machine being able to reach 10,000 feet in about 5 minutes and 20,000 feet in under half an hour. The loading per square foot was also considerably increased; this being rendered possible both by improvement in the design of wing sections and by more scientific construction giving increased strength. It will be remembered that in the machine of the very early period each square foot of surface had only to lift a weight of some 1 1/2 to 2 lbs., which by 1914 had been increased to about 4 lbs. By 1918 aeroplanes habitually had a loading of 8 lbs. or more per square foot of area; which resulted in great increase in speed. Although a speed of 126 miles per hour had been attained by a specially designed racing machine over a short distance in 1914, the average at that period little exceeded, if at all, 100 miles per hour; whereas in 1918 speeds of 130 miles per hour had become a commonplace, and shortly afterwards a speed of over 166 miles an hour was achieved.

In another direction, also, that of size, great developments were made. Before the War a few machines fitted with more than one engine had been built (the first being a triple Gnome-engined biplane built by Messrs Short Bros. at Eastchurch in 1913), but none of large size had been successfully produced, the total weight probably in no case exceeding about 2 tons. In 1916, however, the twin engine Handley-Page biplane was produced, to be followed by others both in this country and abroad, which represented a very great increase in size and, consequently, load-carrying capacity. By the end of the War period several types were in existence weighing a total of 10 tons when fully loaded, of which some 4 tons or more represented 'useful load' available for crew, fuel, and bombs or passengers. This was attained through very careful attention to detailed design, which showed that the material could be employed more efficiently as size increased, and was also due to the fact that a large machine was not liable to be put through the same evolutions as a small machine, and therefore could safely be built with a lower factor of safety. Owing to the fact that a wing section which is adopted for carrying heavy loads usually has also a somewhat low lift to drag ratio, and is not therefore productive of high speed, these machines are not as fast as light scouts; but, nevertheless, they proved themselves capable of achieving speeds of 100 miles an hour or more in some cases; which was faster than the average small machine of 1914.

In one respect the development during the War may perhaps have proved to be somewhat disappointing, as it might have been expected that great improvements would be effected in metal construction, leading almost to the abolition of wooden structures. Although, however, a good deal of experimental work was done which resulted in overcoming at any rate the worst of the difficulties, metal-built machines were little used (except to a certain extent in Germany) chiefly on account of the need for rapid production and the danger of delay resulting from switching over from known and tried methods to experimental types of construction. The Germans constructed some large machines, such as the giant Siemens-Schukhert machine, entirely of metal except for the wing covering, while the Fokker and Junker firms about the time of the Armistice in 1918 both produced monoplanes with very deep all-metal wings (including the covering) which were entirely unstayed externally, depending for their strength on internal bracing. In Great Britain cable bracing gave place to a great extent to 'stream-line wires,' which are steel rods rolled to a more or less oval section, while tie-rods were also extensively used for the internal bracing of the wings. Great developments in the economical use of material were also made in the direction of using built-up main spars for the wings and interplane struts; spars composed of a series of layers (or 'laminations') of different pieces of wood also being used.

Apart from the metallic construction of aeroplanes an enormous amount of work was done in the testing of different steels and light alloys for use in engines, and by the end of the War period a number of aircraft engines were in use of which the pistons and other parts were of such alloys; the chief difficulty having been not so much in the design as in the successful heat-treatment and casting of the metal.

An important development in connection with the inspection and testing of aircraft parts, particularly in the case of metal, was the experimental application of X-ray photography, which showed up latent defects, both in the material and in manufacture, which would otherwise have passed unnoticed. This method was also used to test the penetration of glue into the wood on each side of joints, so giving a measure of the strength; and for the effect of 'doping' the wings, dope being a film (of cellulose acetate dissolved in acetone with other chemicals) applied to the covering of wings and bodies to render the linen taut and weatherproof, besides giving it a smooth surface for the lessening of 'skin friction' when passing rapidly through the air.

An important result of this experimental work was that it in many cases enabled designers to produce aeroplane parts from less costly material than had previously been considered necessary, without impairing the strength. It may be mentioned that it was found undesirable to use welded joints on aircraft in any part where the material is subjectto a tensile or bending load, owing to the danger resulting from bad workmanship causing the material to become brittle--an effect which cannot be discovered except by cutting through the weld, which, of course, involves a test to destruction. Written, as it has been, in August, 1920, it is impossible in this chapter to give any conception of how the developments of War will be applied to commercial aeroplanes, as few truly commercial machines have yet been designed, and even those still show distinct traces of the survival of war mentality. When, however, the inevitable recasting of ideas arrives, it will become evident, whatever the apparent modification in the relative importance of different aspects of design, that enormous advances were made under the impetus of War which have left an indelible mark on progress.

We have, during the seventeen years since aeroplanes first took the air, seen them grow from tentative experimental structures of unknown and unknowable performance to highly scientific products, of which not only the performances (in speed, load-carrying capacity, and climb) are known, but of which the precise strength and degree of stability can be forecast with some accuracy on the drawing board. For the rest, with the future lies--apart from some revolutionary change in fundamental design--the steady development of a now well-tried and well-found engineering structure.

PART III. AEROSTATICS
I. BEGINNINGS

Francesco Lana, with his 'aerial ship,' stands as one of the first great exponents of aerostatics; up to the time of the Montgolfier and Charles balloon experiments, aerostatic and aerodynamic research are so inextricably intermingled that it has been thought well to treat of them as one, and thus the work of Lana, Veranzio and his parachute, Guzman's frauds, and the like, have already been sketched. In connection with Guzman, Hildebrandt states in his Airships Past and Present, a fairly exhaustive treatise on the subject up to 1906, the year of its publication, that there were two inventors--or charlatans--Lorenzo de Guzman and a monk Bartolemeo Laurenzo, the former of whom constructed an unsuccessful airship out of a wooden basket covered with paper, while the latter made certain experiments with a machine of which no description remains. A third de Guzman, some twenty-five years later, announced that he had constructed a flying machine, with which he proposed to fly from a tower to prove his success to the public. The lack of record of any fatal accident overtaking him about that time seems to show that the experiment was not carried out.

Galien, a French monk, published a book L'art de naviguer dans l'air in 1757, in which it was conjectured that the air at high levels was lighter than that immediately over the surface of the earth. Galien proposed to bring down the upper layers of air and with them fill a vessel, which by Archimidean principle would rise through the heavier atmosphere. If one went high enough, said Galien, the air would be two thousand times as light as water, and it would be possible to construct an airship, with this light air as lifting factor, which should be as large as the town of Avignon, and carry four million passengers with their baggage. How this high air was to be obtained is matter for conjecture--Galien seems to have thought in a vicious circle, in which the vessel that must rise to obtain the light air must first be filled with it in order to rise.

Cavendish's discovery of hydrogen in 1776 set men thinking, and soon a certain Doctor Black was suggesting that vessels might be filled with hydrogen, in order that they might rise in the air. Black, however, did not get beyond suggestion; it was Leo Cavallo who first made experiments with hydrogen, beginning with filling soap bubbles, and passing on to bladders and special paper bags. In these latter the gas escaped, and Cavallo was about to try goldbeaters' skin at the time that the Montgolfiers came into the field with their hot air balloon.

Joseph and Stephen Montgolfier, sons of a wealthy French paper manufacturer, carried out many experiments in physics, and Joseph interested himself in the study of aeronautics some time before the first balloon was constructed by the brothers--he is said to have made a parachute descent from the roof of his house as early as 1771, but of this there is no proof. Galien's idea, together with study of the movement of clouds, gave Joseph some hope of achieving aerostation through Galien's schemes, and the first experiments were made by passing steam into a receiver, which, of course, tended to rise--but the rapid condensation of the steam prevented the receiver from more than threatening ascent. The experiments were continued with smoke, which produced only a slightly better effect, and, moreover, the paper bag into which the smoke was induced permitted of escape through its pores; finding this method a failure the brothers desisted until Priestley's work became known to them, and they conceived the use of hydrogen as a lifting factor. Trying this with paper bags, they found that the hydrogen escaped through the pores of the paper.

Their first balloon, made of paper, reverted to the hot-air principle; they lighted a fire of wool and wet straw under the balloon--and as a matter of course the balloon took fire after very little experiment; thereupon they constructed a second, having a capacity of 700 cubic feet, and this rose to a height of over 1,000 feet. Such a success gave them confidence, and they gave their first public exhibition on June 5th, 1783, with a balloon constructed of paper and of a circumference of 112 feet. A fire was lighted under this balloon, which, after rising to a height of 1,000 feet, descended through the cooling of the air inside a matter of ten minutes. At this the Academie des Sciences invited the brothers to conduct experiments in Paris.

The Montgolfiers were undoubtedly first to send up balloons, but other experimenters were not far behind them, and before they could get to Paris in response to their invitation, Charles, a prominent physicist of those days, had constructed a balloon of silk, which he proofed against escape of gas with rubber--the Roberts had just succeeded in dissolving this substance to permit of making a suitable coating for the silk. With a quarter of a ton of sulphuric acid, and half a ton of iron filings and turnings, sufficient hydrogen was generated in four days to fill Charles's balloon, which went up on August 28th, 1783. Although the day was wet, Paris turned out to the number of over 300,000 in the Champs de Mars, and cannon were fired to announce the ascent of the balloon. This, rising very rapidly, disappeared amid the rain clouds, but, probably bursting through no outlet being provided to compensate for the escape of gas, fell soon in the neighbourhood of Paris. Here peasants, ascribing evil supernatural influence to the fall of such a thing from nowhere, went at it with the implements of their craft--forks, hoes, and the like--and maltreated it severely, finally attaching it to a horse's tail and dragging it about until it was mere rag and scrap.

Meanwhile, Joseph Montgolfier, having come to Paris, set about the construction of a balloon out of linen; this was in three diverse sections, the top being a cone 30 feet in depth, the middle a cylinder 42 feet in diameter by 26 feet in depth, and the bottom another cone 20 feet in depth from junction with the cylindrical portion to its point. The balloon was both lined and covered with paper, decorated in blue and gold. Before ever an ascent could be attempted this ambitious balloon was caught in a heavy rainstorm which reduced its paper covering to pulp and tore the linen at its seams, so that a supervening strong wind tore the whole thing to shreds.

Montgolfier's next balloon was spherical, having a capacity of 52,000 cubic feet. It was made from waterproofed linen, and on September 19th, 1783, it made an ascent for the palace courtyard at Versailles, taking up as passengers a cock, a sheep, and a duck. A rent at the top of the balloon caused it to descend within eight minutes, and the duck and sheep were found none the worse for being the first living things to leave the earth in a balloon, but the cock, evidently suffering, was thought to have been affected by the rarefaction of the atmosphere at the tremendous height reached--for at that time the general opinion was that the atmosphere did not extend more than four or five miles above the earth's surface. It transpired later that the sheep had trampled on the cock, causing more solid injury than any that might be inflicted by rarefied air in an eight-minute ascent and descent of a balloon.

For achieving this flight Joseph Montgolfier received from the King of France a pension of of L40, while Stephen was given the order of St Michael, and a patent of nobility was granted to their father. They were made members of the Legion d'Honneur, and a scientific deputation, of which Faujas de Saint-Fond, who had raised the funds with which Charles's hydrogen balloon was constructed, presented to Stephen Montgolfier a gold medal struck in honour of his aerial conquest. Since Joseph appears to have had quite as much share in the success as Stephen, the presentation of the medal to one brother only was in questionable taste, unless it was intended to balance Joseph's pension.

Once aerostation had been proved possible, many people began the construction of small balloons--the wholehole thing was regarded as a matter of spectacles and a form of amusement by the great majority. A certain Baron de Beaumanoir made the first balloon of goldbeaters' skin, this being eighteen inches in diameter, and using hydrogen as a lifting factor. Few people saw any possibilities in aerostation, in spite of the adventures of the duck and sheep and cock; voyages to the moon were talked and written, and there was more of levity than seriousness over ballooning as a rule. The classic retort of Benjamin Franklin stands as an exception to the general rule: asked what was the use of ballooning--'What's the use of a baby?' he countered, and the spirit of that reply brought both the dirigible and the aeroplane to being, later.

The next noteworthy balloon was one by Stephen Montgolfier, designed to take up passengers, and therefore of rather large dimensions, as these things went then. The capacity was 100,000 cubic feet, the depth being 85 feet, and the exterior was very gaily decorated. A short, cylindrical opening was made at the lower extremity, and under this a fire-pan was suspended, above the passenger car of the balloon. On October 15th, 1783, Pilatre de Rozier made the first balloon ascent--but the balloon was held captive, and only allowed to rise to a height of 80 feet. But, a little later in 1783, Rozier secured the honour of making the first ascent in a free balloon, taking up with him the Marquis d'Arlandes. It had been originally intended that two criminals, condemned to death, should risk their lives in the perilous venture, with the prospect of a free pardon if they made a safe descent, but d'Arlandes got the royal consent to accompany Rozier, and the criminals lost their chance. Rozier and d'Arlandes made a voyage lasting for twenty-five minutes, and, on landing, the balloon collapsed with such rapidity as almost to suffocate Rozier, who, however, was dragged out to safety by d'Arlandes. This first aerostatic journey took place on November 21st, 1783.

Some seven months later, on June 4th, 1784, a Madame Thible ascended in a free balloon, reaching a height of 9,000 feet, and making a journey which lasted for forty-five minutes--the great King Gustavus of Sweden witnessed this ascent. France grew used to balloon ascents in the course of a few months, in spite of the brewing of such a storm as might have been calculated to wipe out all but purely political interests. Meanwhile, interest in the new discovery spread across the Channel, and on September 15th, 1784, one Vincent Lunardi made the first balloon voyage in England, starting from the Artillery Ground at Chelsea, with a cat and dog as passengers, and landing in a field in the parish of Standon, near Ware. There is a rather rare book which gives a very detailed account of this first ascent in England, one copy of which is in the library of the Royal Aeronautical Society; the venturesome Lunardi won a greater measure of fame through his exploit than did Cody for his infinitely more courageous and--from a scientific point of view--valuable first aeroplane ascent in this country.

The Montgolfier type of balloon, depending on hot air for its lifting power, was soon realised as having dangerous limitations. There was always a possibility of the balloon catching fire while it was being filled, and on landing there was further danger from the hot pan which kept up the supply of hot air on the voyage--the collapsing balloon fell on the pan, inevitably. The scientist Saussure, observing the filling of the balloons very carefully, ascertained that it was rarefaction of the air which was responsible for the lifting power, and not the heat in itself, and, owing to the rarefaction of the air at normal temperature at great heights above the earth, the limit of ascent for a balloon of the Montgolfier type was estimated by him at under 9,000 feet. Moreover, since the amount of fuel that could be carried for maintaining the heat of the balloon after inflation was subject to definite limits, prescribed by the carrying capacity of the balloon, the duration of the journey was necessarily limited just as strictly.

These considerations tended to turn the minds of those interested in aerostation to consideration of the hydrogen balloon evolved by Professor Charles. Certain improvements had been made by Charles since his first construction; he employed rubber-coated silk in the construction of a balloon of 30 feet diameter, and provided a net for distributing the pressure uniformly over the surface of the envelope; this net covered the top half of the balloon, and from its lower edge dependent ropes hung to join on a wooden ring, from which the car of the balloon was suspended-- apart from the extension of the net so as to cover in the whole of the envelope, the spherical balloon of to-day is virtually identical with that of Charles in its method of construction. He introduced the valve at the top of the balloon, by which escape of gas could be controlled, operating his valve by means of ropes which depended to the car of the balloon, and he also inserted a tube, of about 7 inches diameter, at the bottom of the balloon, not only for purposes of inflation, but also to provide a means of escape for gas in case of expansion due to atmospheric conditions.

Sulphuric acid and iron filings were used by Charles for filling his balloon, which required three days and three nights for the generation of its 14,000 cubic feet of hydrogen gas. The inflation was completed on December 1st, 1783, and the fittings carried included a barometer and a grapnel form of anchor. In addition to this, Charles provided the first 'ballon sonde' in the form of a small pilot balloon which he handed to Montgolfier to launch before his own ascent, in order to determine the direction and velocity of the wind. It was a graceful compliment to his rival, and indicated that, although they were both working to the one end, their rivalry was not a matter of bitterness.

Ascending on December 1st, 1783, Charles took with him one of the brothers Robert, and with him made the record journey up to that date, covering a period of three and three-quarter hours, in which time they journeyed some forty miles. Robert then landed, and Charles ascended again alone, reaching such a height as to feel the effects of the rarefaction of the air, this very largely due to the rapidity of his ascent. Opening the valve at the top of the balloon, he descended thirty-five minutes after leaving Robert behind, and came to earth a few miles from the point of the first descent. His discomfort over the rapid ascent was mainly due to the fact that, when Robert landed, he forgot to compensate for the reduction of weight by taking in further ballast, but the ascent proved the value of the tube at the bottom of the balloon envelope, for the gas escaped very rapidly in that second ascent, and, but for the tube, the balloon must inevitably have burst in the air, with fatal results for Charles.

As in the case of aeroplane flight, as soon as the balloon was proved practicable the flight across the English Channel was talked of, and Rozier, who had the honour of the first flight, announced his intention of being first to cross. But Blanchard, who had an idea for a 'flying car,' anticipated him, and made a start from Dover on January 7th, 1785, taking with him an American doctor named Jeffries. Blanchard fitted out his craft for the journey very thoroughly, taking provisions, oars, and even wings, for propulsion in case of need. He took so much, in fact, that as soon as the balloon lifted clear of the ground the whole of the ballast had to be jettisoned, lest the balloon should drop into the sea. Half-way across the Channel the sinking of the balloon warned Blanchard that he had to part with more than ballast to accomplish the journey, and all the equipment went, together with certain books and papers that were on board the car. The balloon looked perilously like collapsing, and both Blanchard and Jeffries began to undress in order further to lighten their craft--Jeffries even proposed a heroic dive to save the situation, but suddenly the balloon rose sufficiently to clear the French coast, and the two voyagers landed at a point near Calais in the Forest of Gaines, where a marble column was subsequently erected to commemorate the great feat.

Rozier, although not first across, determined to be second, and for that purpose he constructed a balloon which was to owe its buoyancy to a combination of the hydrogen and hot air principles. There was a spherical hydrogen balloon above, and beneath it a cylindrical container which could be filled with hot air, thus compensating for the leakage of gas from the hydrogen portion of the balloon--regulating the heat of his fire, he thought, would give him perfect control in the matter of ascending and descending.

On July 6th, 1785, a favourable breeze gave Rozier his opportunity of starting from the French coast, and with a passenger aboard he cast off in his balloon, which he had named the 'Aero-Montgolfiere.' There was a rapid rise at first, and then for a time the balloon remained stationary over the land, after which a cloud suddenly appeared round the balloon, denoting that an explosion had taken place. Both Rozier and his companion were killed in the fall, so that he, first to leave the earth by balloon, was also first victim to the art of aerostation.

There followed, naturally, a lull in the enthusiasm with which ballooning had been taken up, so far as France was concerned. In Italy, however, Count Zambeccari took up hot-air ballooning, using a spirit lamp to give him buoyancy, and on the first occasion when the balloon car was set on fire Zambeccari let down his passenger by means of the anchor rope, and managed to extinguish the fire while in the air. This reduced the buoyancy of the balloon to such an extent that it fell into the Adriatic and was totally wrecked, Zambeccari being rescued by fishermen. He continued to experiment up to 1812, when he attempted to ascend at Bologna; the spirit in his lamp was upset by the collision of the car with a tree, and the car was again set on fire. Zambeccari jumped from the car when it was over fifty feet above level ground, and was killed. With him the Rozier type of balloon, combining the hydrogen and hot air principles, disappeared; the combination was obviously too dangerous to be practical.

The brothers Robert were first to note how the heat of the sun acted on the gases within a balloon envelope, and it has since been ascertained that sun rays will heat the gas in a balloon to as much as 80 degrees Fahrenheit greater temperature than the surrounding atmosphere; hydrogen, being less affected by change of temperature than coal gas, is the most suitable filling element, and coal gas comes next as the medium of buoyancy. This for the free and non-navigable balloon, though for the airship, carrying means of combustion, and in military work liable to ignition by explosives, the gas helium seems likely to replace hydrogen, being non-combustible.

In spite of the development of the dirigible airship, there remains work for the free, spherical type of balloon in the scientific field. Blanchard's companion on the first Channel crossing by balloon, Dr Jeffries, was the first balloonist to ascend for purely scientific purposes; as early as 1784 he made an ascent to a height of 9,000 feet, and observed a fall in temperature of from degrees--at the level of London, where he began his ascent--to 29 degrees at the maximum height reached. He took up an electrometer, a hydrometer, a compass, a thermometer, and a Toricelli barometer, together with bottles of water, in order to collect samples of the air at different heights. In 1785 he made a second ascent, when trigonometrical observations of the height of the balloon were made from the French coast, giving an altitude of

4,800 feet.

The matter was taken up on its scientific side very early in America, experiments in Philadelphia being almost simultaneous with those of the Montgolfiers in France. The flight of Rozier and d'Arlandes inspired two members of the Philadelphia Philosophical Academy to construct a balloon or series of balloons of their own design; they made a machine which consisted of no less than 47 small hydrogen balloons attached to a wicker car, and made certain preliminary trials, using animals as passengers. This was followed by a captive ascent with a man as passenger, and eventually by the first free ascent in America, which was undertaken by one James Wilcox, a carpenter, on December 28th, 1783. Wilcox, fearful of falling into a river, attempted to regulate his landing by cutting slits in some of the supporting balloons, which was the method adopted for regulating ascent or descent in this machine. He first cut three, and then, finding that the effect produced was not sufficient, cut three more, and then another five--eleven out of the forty-seven. The result was so swift a descent that he dislocated his wrist on landing.

A NOTE ON BALLONETS OR AIR BAGS.

Meusnier, toward the end of the eighteenth century, was first to conceive the idea of compensating for the loss of gas due to expansion by fitting to the interior of a free balloon a ballonet, or air bag, which could be pumped full of air so as to retain the shape and rigidity of the envelope.

The ballonet became particularly valuable as soon as airship construction became general, and it was in the course of advance in Astra Torres design that the project was introduced of using the ballonets in order to give inclination from the horizontal. In the earlier Astra Torres, trimming was accomplished by moving the car fore and aft--this in itself was an advance on the separate 'sliding weigh' principle--and this was the method followed in the Astra Torres bought by the British Government from France in 1912 for training airship pilots. Subsequently, the two ballonets fitted inside the envelope were made to serve for trimming by the extent of their inflation, and this method of securing inclination proved the best until exterior rudders, and greater engine power, supplanted it, as in the Zeppelin and, in fact, all rigid types.

In the kite balloon, the ballonet serves the purpose of a rudder, filling itself through the opening being kept pointed toward the wind--there is an ingenious type of air scoop with non-return valve which assures perfect inflation. In the S.S. type of airship, two ballonets are provided, the supply of air being taken from the propeller draught by a slanting aluminium tube to the underside of the envelope, where it meets a longitudinal fabric hose which connects the two ballonet air inlets. In this hose the non-return air valves, known as 'crab-pots,' are fitted, on either side of the junction with the air-scoop. Two automatic air valves, one for each ballonet, are fitted in the underside of the envelope, and, as the air pressure tends to open these instead of keeping them shut, the spring of the valve is set inside the envelope. Each spring is set to open at a pressure of 25 to 28 mm.

II. THE FIRST DIRIGIBLES

Having got off the earth, the very early balloonists set about the task of finding a means of navigating the air but, lacking steam or other accessory power to human muscle, they failed to solve the problem. Joseph Montgolfier speedily exploded the idea of propelling a balloon either by means of oars or sails, pointing out that even in a dead calm a speed of five miles an hour would be the limit achieved. Still, sailing balloons were constructed, even up to the time of Andree, the explorer, who proposed to retard the speed of the balloon by ropes dragging on the ground, and then to spread a sail which should catch the wind and permit of deviation of the course. It has been proved that slight divergences from the course of the wind can be obtained by this means, but no real navigation of the air could be thus accomplished.

Professor Wellner, of Brunn, brought up the idea of a sailing balloon in more practical fashion in 1883. He observed that surfaces inclined to the horizontal have a slight lateral motion in rising and falling, and deduced that by alternate lowering and raising of such surfaces he would be able to navigate the air, regulating ascent and descent by increasing or decreasing the temperature of his buoyant medium in the balloon. He calculated that a balloon, 50 feet in diameter and 150 feet in length, with a vertical surface in front and a horizontal surface behind, might be navigated at a speed of ten miles per hour, and in actual tests at Brunn he proved that a single rise and fall moved the balloon three miles against the wind. His ideas were further developed by Lebaudy in the construction of the early French dirigibles.

According to Hildebrandt[11], the first sailing balloon was built in 1784 by Guyot, who made his balloon egg-shaped, with the smaller end at the back and the longer axis horizontal; oars were intended to propel the craft, and naturally it was a failure. Carra proposed the use of paddle wheels, a step in the right direction, by mounting them on the sides of the car, but the improvement was only slight. Guyton de Morveau, entrusted by the Academy of Dijon with the building of a sailing balloon, first used a vertical rudder at the rear end of his construction--it survives in the modern dirigible. His construction included sails and oars, but, lacking steam or other than human propulsive power, the airship was a failure equally with Guyot's.

Two priests, Miollan and Janinet, proposed to drive balloons through the air by the forcible expulsion of the hot air in the envelope from the rear of the balloon. An opening was made about half-way up the envelope, through which the hot air was to escape, buoyancy being maintained by a pan of combustibles in the car. Unfortunately, this development of the Montgolfier type never got a trial, for those who were to be spectators of the first flight grew exasperated at successive delays, and in the end, thinking that the balloon would never rise, they destroyed it.

Meusnier, a French general, first conceived the idea of compensating for loss of gas by carrying an air bag inside the balloon, in order to maintain the full expansion of the envelope. The brothers Robert constructed the first balloon in which this was tried and placed the air bag near the neck of the balloon which was intended to be driven by oars, and steered by a rudder. A violent swirl of wind which was encountered on the first ascent tore away the oars and rudder and broke the ropes which held the air bag in position; the bag fell into the opening of the neck and stopped it up, preventing the escape of gas under expansion. The Duc de Chartres, who was aboard, realised the extreme danger of the envelope bursting as the balloon ascended, and at 16,000 feet he thrust a staff through the envelope--another account says that he slit it with his sword--and thus prevented disaster. The descent after this rip in the fabric was swift, but the passengers got off without injury in the landing.

Meusnier, experimenting in various ways, experimented with regard to the

---
11 Airships Past and Present.

resistance offered by various shapes to the air, and found that an elliptical shape was best; he proposed to make the car boat--shaped, in order further to decrease the resistance, and he advocated an entirely rigid connection between the car and the body of the balloon, as indispensable to a dirigible[12]. He suggested using three propellers, which were to be driven by hand by means of pulleys, and calculated that a crew of eighty would be required to furnish sufficient motive power. Horizontal fins were to be used to assure stability, and Meusnier thoroughly investigated the pressures exerted by gases, in order to ascertain the stresses to which the envelope would be subjected. More important still, he went into detail with regard to the use of air bags, in order to retain the shape of the balloon under varying pressures of gas due to expansion and consequent losses; he proposed two separate envelopes, the inner one containing gas, and the space between it and the outer one being filled with air. Further, by compressing the air inside the air bag, the rate of ascent or descent could be regulated. Lebaudy, acting on this principle, found it possible to pump air at the rate of 35 cubic feet per second, thus making good loss of ballast which had to be thrown overboard.

Meusnier's balloon, of course, was never constructed, but his ideas have been of value to aerostation up to the present time. His career ended in the revolutionary army in 1793, when he was killed in the fighting before Mayence, and the King of Prussia ordered all firing to cease until Meusnier had been buried. No other genius came forward to carry on his work, and it was realised that human muscle could not drive a balloon with certainty through the air; experiment in this direction was abandoned for nearly sixty years, until in 1852 Giffard brought the first practicable power-driven dirigible to being.

Giffard, inventor of the steam injector, had already made balloon ascents when he turned to aeronautical propulsion, and constructed a steam engine of 5 horsepower with a weight of only 100 lbs.--a great achievement for his day. Having got his engine, he set about making the balloon which it was to drive; this he built with the aid of two other enthusiasts, diverging from Meusnier's ideas by making the ends pointed, and keeping the body narrowed from Meusnier's ellipse to a shape more resembling a rather fat cigar. The length was 144 feet, and the greatest diameter only 40 feet, while the capacity was 88,000 cubic feet. A net which covered

12 Hildebrandt.

the envelope of the balloon supported a spar, 66 feet in length, at the end of which a triangular sail was placed vertically to act as rudder. The car, slung 20 feet below the spar, carried the engine and propeller. Engine and boiler together weighed 350 lbs., and drove the 11 foot propeller at 110 revolutions per minute.

As precaution against explosion, Giffard arranged wire gauze in front of the stoke-hole of his boiler, and provided an exhaust pipe which discharged the waste gases from the engine in a downward direction. With this first dirigible he attained to a speed of between 6 and 8 feet per second, thus proving that the propulsion of a balloon was a possibility, now that steam had come to supplement human effort.

Three years later he built a second dirigible, reducing the diameter and increasing the length of the gas envelope, with a view to reducing air resistance. The length of this was 230 feet, the diameter only 33 feet, and the capacity was 113,000 cubic feet, while the upper part of the envelope, to which the covering net was attached, was specially covered to ensure a stiffening effect. The car of this dirigible was dropped rather lower than that of the first machine, in order to provide more thoroughly against the danger of explosions. Giffard, with a companion named Yon as passenger, took a trial trip on this vessel, and made a journey against the wind, though slowly. In commencing to descend, the nose of the envelope tilted upwards, and the weight of the car and its contents caused the net to slip, so that just before the dirigible reached the ground, the envelope burst. Both Giffard and his companion escaped with very slight injuries.

Plans were immediately made for the construction of a third dirigible, which was to be 1,970 feet in length, 98 feet in extreme diameter, and to have a capacity of 7,800,000 cubic feet of gas. The engine of this giant was to have weighed 30 tons, and with it Giffard expected to attain a speed of 40 miles per hour. Cost prevented the scheme being carried out, and Giffard went on designing small steam engines until his invention of the steam injector gave him the funds to turn to dirigibles again. He built a captive balloon for the great exhibition in London in 1868, at a cost of nearly L30,000, and designed a dirigible balloon which was to have held a million and three quarters cubic feet of gas, carry two boilers, and cost about L40,000. The plans were thoroughly worked out, down to the last detail, but the dirigible was never constructed. Giffard went blind, and died in 1882--he stands as the great pioneer of dirigible construction, more on the strength of the two vessels which he actually built than on that of the ambitious later conceptions of his brain.

In 1872 Dupuy de Lome, commissioned by the French government, built a dirigible which he proposed to drive by man-power--it was anticipated that the vessel would be of use in the siege of Paris, but it was not actually tested till after the conclusion of the war. The length of this vessel was 118 feet, its greatest diameter 49 feet, the ends being pointed, and the motive power was by a propeller which was revolved by the efforts of eight men. The vessel attained to about the same speed as Giffard's steam-driven airship; it was capable of carrying fourteen men, who, apart from these engaged in driving the propeller, had to manipulate the pumps which controlled the air bags inside the gas envelope.

In the same year Paul Haenlein, working in Vienna, produced an airship which was a direct forerunner of the Lebaudy type, 164 feet in length, 30 feet greatest diameter, and with a cubic capacity of 85,000 feet. Semi-rigidity was attained by placing the car as close to the envelope as possible, suspending it by crossed ropes, and the motive power was a gas engine of the Lenoir type, having four horizontal cylinders, and giving about 5 horse-power with a consumption of about 250 cubic feet of gas per hour. This gas was sucked from the envelope of the balloon, which was kept fully inflated by pumping in compensating air to the air bags inside the main envelope. A propeller, 15 feet in diameter, was driven by the Lenoir engine at 40 revolutions per minute. This was the first instance of the use of an internal combustion engine in connection with aeronautical experiments.

The envelope of this dirigible was rendered airtight by means of internal rubber coating, with a thinner film on the outside. Coal gas, used for inflation, formed a suitable fuel for the engine, but limited the height to which the dirigible could ascend. Such trials as were made were carried out with the dirigible held captive, and a speed of 15 feet per second was attained. Full experiment was prevented through funds running low, but Haenlein's work constituted a distinct advance on all that had been done previously.

Two brothers, Albert and Gaston Tissandier, were next to enter the field of dirigible construction; they had experimented with balloons during the Franc-Prussian War, and had attempted to get into Paris by balloon during the siege, but it was not until 1882 that they produced their dirigible.

This was 92 feet in length and 32 feet in greatest diameter, with a cubic capacity of 37,500 feet, and the fabric used was varnished cambric. The car was made of bamboo rods, and in addition to its crew of three, it carried a Siemens dynamo, with 24 bichromate cells, each of which weighed 17 lbs. The motor gave out 1 1/2 horse-power, which was sufficient to drive the vessel at a speed of up to 10 feet per second. This was not so good as Haenlein's previous attempt and, after L2,000 had been spent, the Tissandier abandoned their experiments, since a 5-mile breeze was sufficient to nullify the power of the motor.

Renard, a French officer who had studied the problem of dirigible construction since 1878, associated himself first with a brother officer named La Haye, and subsequently with another officer, Krebs, in the construction of the second dirigible to be electrically-propelled. La Haye first approached Colonel Laussedat, in charge of the Engineers of the French Army, with a view to obtaining funds, but was refused, in consequence of the practical failure of all experiments since 1870. Renard, with whom Krebs had now associated himself, thereupon went to Gambetta, and succeeded in getting a promise of a grant of L8,000 for the work; with this promise Renard and Krebs set to work.

They built their airship in torpedo shape, 165 feet in length, and of just over 27 feet greatest diameter--the greatest diameter was at the front, and the cubic capacity was 66,000 feet. The car itself was 108 feet in length, and 4 1/2 feet broad, covered with silk over the bamboo framework. The 23 foot diameter propeller was of wood, and was driven by an electric motor connected to an accumulator, and yielding 8.5 horsepower. The sweep of the propeller, which might have brought it in contact with the ground in landing, was counteracted by rendering it possible to raise the axis on which the blades were mounted, and a guide rope was used to obviate damage altogether, in case of rapid descent. There was also a 'sliding weight' which was movable to any required position to shift the centre of gravity as desired. Altogether, with passengers and ballast aboard, the craft weighed two tons.

In the afternoon of August 8th, 1884, Renard and Krebs ascended in the dirigible--which they had named 'La France,' from the military ballooning ground at Chalais-Meudon, making a circular flight of about five miles, the latter part of which was in the face of a slight wind. They found that the vessel answered well to her rudder, and the five-mile flight was made successfully in a period of 23 minutes. Subsequent experimental flights determined that the air speed of the dirigible was no less than 14 1/2 miles per hour, by far the best that had so far been accomplished in dirigible flight. Seven flights in all were made, and of these five were completely successful, the dirigible returning to its starting point with no difficulty. On the other two flights it had to be towed back.

Renard attempted to repeat his construction on a larger scale, but funds would not permit, and the type was abandoned; the motive power was not sufficient to permit of more than short flights, and even to the present time electric motors, with their necessary accumulators, are far too cumbrous to compete with the self-contained internal combustion engine. France had to wait for the Lebaudy brothers, just as Germany had to wait for Zeppelin and Parseval.

Two German experimenters, Baumgarten and Wolfert, fitted a Daimler motor to a dirigible balloon which made its first ascent at Leipzig in 1880. This vessel had three cars, and placing a passenger in one of the outer cars[13] distributed the load unevenly, so that the whole vessel tilted over and crashed to the earth, the occupants luckily escaping without injury. After Baumgarten's death, Wolfert determined to carry on with his experiments, and, having achieved a certain measure of success, he announced an ascent to take place on the Tempelhofer Field, near Berlin, on June 12th, 1897. The vessel, travelling with the wind, reached a height of 600 feet, when the exhaust of the motor communicated flame to the envelope of the balloon, and Wolfert, together with a passenger he carried, was either killed by the fall or burnt to death on the ground. Giffard had taken special precautions to avoid an accident of this nature, and Wolfert, failing to observe equal care, paid the full penalty.

Platz, a German soldier, attempting an ascent on the Tempelhofer Field in the Schwartz airship in 1897, merely proved the dirigible a failure. The vessel was of aluminium, 0.008 inch in thickness, strengthened by an aluminium lattice work; the motor was two-cylindered petrol-driven; at the first trial the metal developed such leaks that the vessel came to the ground within four miles of its starting point. Platz, who was aboard alone as crew, succeeded in escaping by jumping clear before the car touched earth, but the shock of alighting broke up the balloon, and a following high wind completed the work of full destruction. A second account says that Platz, finding the propellers insufficient to drive the vessel against the wind, opened the valve and descended too rapidly.

13 Hildebrandt.

The envelope of this dirigible was 156 feet in length, and the method of filling was that of pushing in bags, fill them with gas, and then pulling them to pieces and tearing them out of the body of the balloon. A second contemplated method of filling was by placing a linen envelope inside the aluminium casing, blowing it out with air, and then admitting the gas between the linen and the aluminium outer casing. This would compress the air out of the linen envelope, which was to be withdrawn when the aluminium casing had been completely filled with gas.

All this, however, assumes that the Schwartz type--the first rigid dirigible, by the way--would prove successful. As it proved a failure on the first trial, the problem of filling it did not arise again.

By this time Zeppelin, retired from the German army, had begun to devote himself to the study of dirigible construction, and, a year after Schwartz had made his experiment and had failed, he got together sufficient funds for the formation of a limitedliability company, and started on the construction of the first of his series of airships. The age of tentative experiment was over, and, forerunner of the success of the heavier-than-air type of flying machine, successful dirigible flight was accomplished by Zeppelin in Germany, and by Santos-Dumont in France.

III. SANTOS-DUMONT

A Brazilian by birth, Santos-Dumont began in Paris in the year 1898 to make history, which he subsequently wrote. His book, My Airships, is a record of his eight years of work on lighter-than-air machines, a period in which he constructed no less than fourteen dirigible balloons, beginning with a cubic capacity of 6,350 feet, and an engine of 3 horse-power, and rising to a cubic capacity of 71,000 feet on the tenth dirigible he constructed, and an engine of 60 horse-power, which was fitted to the seventh machine in order of construction, the one which he built after winning the Deutsch Prize.

The student of dirigible construction is recommended to Santos-Dumont's own book not only as a full record of his work, but also as one of the best stories of aerial navigation that has ever been written. Throughout all his experiments, he adhered to the non-rigid type; his first dirigible made its first flight on September 18th, 1898, starting from the Jardin d'Acclimatation to the west of Paris; he calculated that his 3 horse-power engine would yield sufficient power to enable him to steer clear of the trees with which the starting-point was surrounded, but, yielding to the advice of professional aeronauts who were present, with regard to the placing of the dirigible for his start, he tore the envelope against the trees. Two days later, having repaired the balloon, he made an ascent of 1,300 feet. In descending, the hydrogen left in the balloon contracted, and Santos-Dumont narrowly escaped a serious accident in coming to the ground.

His second machine, built in the early spring of 1899, held over 7,000 cubic feet of gas and gave a further 44 lbs. of ascensional force. The balloon envelope was very long and very narrow; the first attempt at flight was made in wind and rain, and the weather caused sufficient contraction of the hydrogen for a wind gust to double the machine up and toss it into the trees near its starting-point. The inventor immediately set about the construction of 'Santos-Dumont No. 3,' on which he made a number of successful flights, beginning on November 13th, 1899. On the last of his flights, he lost the rudder of the machine and made a fortunate landing at Ivry. He did not repair the balloon, considering it too clumsy in form and its motor too small. Consequently No. 4 was constructed, being finished on the 1st, August, 1900. It had a cubic capacity of 14,800 feet, a length of 129 feet and greatest diameter of 16.7 feet, the power plant being a 7 horse-power Buchet motor. Santos-Dumont sat on a bicycle saddle fixed to the long bar suspended under the machine, which also supported motor propeller, ballast; and fuel. The experiment of placing the propeller at the stem instead of at the stern was tried, and the motor gave it a speed of 100 revolutions per minute. Professor Langley witnessed the trials of the machine, which proved before the members of the International Congress of Aeronautics, on September 19th, that it was capable of holding its own against a strong wind.

Finding that the cords with which his dirigible balloon cars were suspended offered almost as much resistance to the air as did the balloon itself, Santos-Dumont substituted piano wire and found that the alteration constituted greater progress than many a more showy device. He altered the shape and size of his No. 4 to a certain extent and fitted a motor of 12 horse-power. Gravity was controlled by shifting weights worked by a cord; rudder and propeller were both placed at the stern. In Santos-Dumont's book there is a certain amount of confusion between the No. 4

and No. 5 airships, until he explains that 'No. 5' is the reconstructed 'No. 4.' It was with No. 5 that he won the Encouragement Prize presented by the Scientific Commission of the Paris Aero Club. This he devoted to the first aeronaut who between May and October of 1900 should start from St Cloud, round the Eiffel Tower, and return. If not won in that year, the prize was to remain open the following year from May 1st to October 1st, and so on annually until won. This was a simplification of the conditions of the Deutsch Prize itself, the winning of which involved a journey of 11 kilometres in 30 minutes.

The Santos-Dumont No. 5, which was in reality the modified No. 4 with new keel, motor, and propeller, did the course of the Deutsch Prize, but with it Santos-Dumont made no attempt to win the prize until July of 1901, when he completed the course in 40 minutes, but tore his balloon in landing. On the 8th August, with his balloon leaking, he made a second attempt, and narrowly escaped disaster, the airship being entirely wrecked. Thereupon he built No. 6 with a cubic capacity of 22,239 feet and a lifting power of 1,518 lbs.

With this machine he won the Deutsch Prize on October 19th, 1901, starting with the disadvantage of a side wind of 20 feet per second. He reached the Eiffel Tower in 9 minutes and, through miscalculating his turn, only just missed colliding with it. He got No. 6 under control again and succeeded in getting back to his starting-point in 29 1/2 minutes, thus winning the 125,000 francs which constituted the Deutsch Prize, together with a similar sum granted to him by the Brazilian Government for the exploit. The greater part of this money was given by Santos-Dumont to charities.

He went on building after this until he had made fourteen non-rigid dirigibles; of these No. 12 was placed at the disposal of the military authorities, while the rest, except for one that was sold to an American and made only one trip, were matters of experiment for their maker. His conclusions from his experiments may be gathered from his own work:--

'On Friday, 31st July, 1903, Commandant Hirschauer and Lieutenant-Colonel Bourdeaux spent the afternoon with me at my airship station at Neuilly St James, where I had my three newest airships--the racing 'No. 7,' the omnibus 'No. 10,' and the runabout 'No. 9'--ready for their study. Briefly, I may say that the opinions expressed by the representatives of the Minister of War were so unreservedly favourable that a practical test of a novel character was decided to be made. Should the airship chosen pass successfully through it the result will be conclusive of its military value.

'Now that these particular experiments are leaving my exclusively private control I will say no more of them than what has been already published in the French press. The test will probably consist of an attempt to enter one of the French frontier towns, such as Belfort or Nancy, on the same day that the airship leaves Paris. It will not, of course, be necessary to make the whole journey in the airship. A military railway wagon may be assigned to carry it, with its balloon uninflated, with tubes of hydrogen to fill it, and with all the necessary machinery and instruments arranged beside it. At some station a short distance from the town to be entered the wagon may be uncoupled from the train, and a sufficient number of soldiers accompanying the officers will unload the airship and its appliances, transport the whole to the nearest open space, and at once begin inflating the balloon. Within two hours from quitting the train the airship may be ready for its flight to the interior of the technically-besieged town.

'Such may be the outline of the task--a task presented imperiously to French balloonists by the events of 1870-1, and which all the devotion and science of the Tissandier brothers failed to accomplish. To-day the problem may be set with better hope of success. All the essential difficulties may be revived by the marking out of a hostile zone around the town that must be entered; from beyond the outer edge of this zone, then, the airship will rise and take its flight--across it.

'Will the airship be able to rise out of rifle range? I have always been the first to insist that the normal place of the airship is in low altitudes, and I shall have written this book to little purpose if I have not shown the reader the real dangers attending any brusque vertical mounting to considerable heights. For this we have the terrible Severo accident before our eyes. In particular, I have expressed astonishment at hearing of experimenters rising to these altitudes without adequate purpose in their early stages of experience with dirigible balloons. All this is very different, however, from a reasoned, cautious mounting, whose necessity has been foreseen and prepared for.'

Probably owing to the fact that his engines were not of sufficient power, Santos-Dumont cannot be said to have solved the problem of the military airship, although the French Government bought one of his vessels. At the same time, he accomplished much in furthering and inciting experiment with dirigible airships, and he will always rank high among the pioneers of aerostation. His experiments might have gone further had not the Wright brothers' success in America and French interest in the problem of the heavier-than-air machine turned him from the study of dirigibles to that of the aeroplane, in which also he takes high rank among the pioneers, leaving the construction of a successful military dirigible to such men as the Lebaudy brothers, Major Parseval, and Zeppelin.

IV. THE MILITARY DIRIGIBLE

Although French and German experiment in connection with the production of an airship which should be suitable for military purposes proceeded side by side, it is necessary to outline the development in the two countries separately, owing to the differing character of the work carried out. So far as France is concerned, experiment began with the Lebaudy brothers, originally sugar refiners, who turned their energies to airship construction in 1899. Three years of work went to the production of their first vessel, which was launched in 1902, having been constructed by them together with a balloon manufacturer named Surcouf and an engineer, Julliot. The Lebaudy airships were what is known as semi-rigids, having a spar which ran practically the full length of the gas bag to which it was attached in such a way as to distribute the load evenly. The car was suspended from the spar, at the rear end of which both horizontal and vertical rudders were fixed, whilst stabilising fins were provided at the stern of the gas envelope itself. The first of the Lebaudy vessels was named the 'Jaune'; its length was 183 feet and its maximum diameter 30 feet, while the cubic capacity was 80,000 feet. The power unit was a 40 horse-power Daimler motor, driving two propellers and giving a maximum speed of 26 miles per hour. This vessel made 29 trips, the last of which took place in November, 1902, when the airship was wrecked through collision with a tree.

The second airship of Lebaudy construction was 7 feet longer than the first, and had a capacity of 94,000 cubic feet of gas with a triple air bag of 17,500 cubic feet to compensate for loss of gas; this latter was kept inflated by a rotary fan. The vessel was eventually taken over by the French Government and may be counted the first dirigible airship considered fit on its tests for military service.

Later vessels of the Lebaudy type were the 'Patrie' and 'Republique,' in which both size and method of construction surpassed those of the two first attempts. The 'Patrie' was fitted with a 60 horse-power engine which gave a speed of 28 miles an hour, while the vessel had a radius of 280 miles, carrying a crew of nine. In the winter of 1907 the 'Patrie' was anchored at Verdun, and encountered a gale which broke her hold on her mooring-ropes. She drifted derelict westward across France, the Channel, and the British Isles, and was lost in the Atlantic.

The 'Republique' had an 80 horse-power motor, which, however, only gave her the same speed as the 'Patrie.' She was launched in July, 1908, and within three months came to an end which constituted a tragedy for France. A propeller burst while the vessel was in the air, and one blade, flying toward the envelope, tore in it a great gash; the airship crashed to earth, and the two officers and two non-commissioned officers who were in the car were instantaneously killed.

The Clement Bayard, and subsequently the Astra-Torres, non-rigids, followed on the early Lebaudys and carried French dirigible construction up to 1912. The Clement Bayard was a simple non-rigid having four lobes at the stern end to assist stability. These were found to retard the speed of the airship, which in the second and more successful construction was driven by a Clement Bayard motor of 100 horse-power at a speed of 30 miles an hour. On August 23rd, 1909, while being tried for acceptance by the military authorities, this vessel achieved a record by flying at a height of 5,000 feet for two hours. The Astra-Torres non-rigids were designed by a Spaniard, Senor Torres, and built by the Astra Company. The envelope was of trefoil shape, this being due to the interior rigging from the suspension band; the exterior appearance is that of two lobes side by side, overlaid by a third. The interior rigging, which was adopted with a view to decreasing air resistance, supports a low-hung car from the centre of the envelope; steering is accomplished by means of horizontal planes fixed on the envelope at the stern, and vertical planes depending beneath the envelope, also at the stern end.

One of the most successful of French pre-war dirigibles was a Clement Bayard built in 1912. In this twin propellers were placed at the front and horizontal and vertical rudders in a sort of box formation under the envelope at the stern. The envelope was stream-lined, while the car of the machine was placed well forward with horizontal controlling planes above it and immediately behind the propellers. This airship, which was named 'Dupuy de Lome,' may be ranked as about the most successful non-rigid dirigible constructed prior to the War.

Experiments with non-rigids in Germany was mainly carried on by Major Parseval, who produced his first vessel in 1906. The main feature of this airship consisted in variation in length of the suspension cables at the will of the operator, so that the envelope could be given an upward tilt while the car remained horizontal in order to give the vessel greater efficiency in climbing. In this machine, the propeller was placed above and forward of the car, and the controlling planes were fixed directly to the envelope near the forward end. A second vessel differed from the first mainly in the matter of its larger size, variable suspension being again employed, together with a similar method of control. The vessel was moderately successful, and under Major Parseval's direction a third was constructed for passenger carrying, with two engines of 120 horsepower, each driving propellers of 13 feet diameter. This was the most successful of the early German dirigibles; it made a number of voyages with a dozen passengers in addition to its crew, as well as proving its value for military purposes by use as a scout machine in manoeuvres. Later Parsevals were constructed of stream-line form, about 300 feet in length, and with engines sufficiently powerful to give them speeds up to 50 miles an hour.

Major Von Gross, commander of a Balloon Battalion, produced semi-rigid dirigibles from 1907 onward. The second of these, driven by two 75 horse-power Daimler motors, was capable of a speed of 27 miles an hour; in September of 1908 she made a trip from and back to Berlin which lasted 13 hours, in which period she covered 176 miles with four passengers and reached a height of 4,000 feet. Her successor, launched in April of 1909, carried a wireless installation, and the next to this, driven by four motors of 75 horse-power each, reached a speed of 45 miles an hour. As this vessel was constructed for military purposes, very few details either of its speed or method of construction were made public.

Practically all these vessels were discounted by the work of Ferdinand von Zeppelin, who set out from the first with the idea of constructing a rigid dirigible. Beginning in 1898, he built a balloon on an aluminium framework covered with linen and silk, and divided into interior compartments holding linen bags which were capable of containing nearly 400,000 cubic feet of hydrogen. The total length of this first Zeppelin airship was 420 feet and the diameter 38 feet. Two cars were rigidly attached to the envelope, each carrying a 16 horse-power motor, driving propellers which were rigidly connected to the aluminium framework of the balloon. Vertical and horizontal screws were used for lifting and forward driving and a sliding weight was used to raise or lower the stem of the vessel out of the horizontal in order to rise or descend without altering the load by loss of ballast or the lift by loss of gas.

The first trial of this vessel was made in July of 1900, and was singularly unfortunate. The winch by which the sliding weight was operated broke, and the balloon was so bent that the working of the propellers was interfered with, as was the steering. A speed of 13 feet per second was attained, but on descending, the airship ran against some piles and was further damaged. Repairs were completed by the end of September, 1900, and on a second trial flight made on October 21st a speed of 30 feet per second was reached.

Zeppelin was far from satisfied with the performance of this vessel, and he therefore set about collecting funds for the construction of a second, which was completed in 1905. By this time the internal combustion engine had been greatly improved, and without any increase of weight, Zeppelin was able to instal two motors of 85 horse-power each. The total capacity was 367,000 cubic feet of hydrogen, carried in 16 gas bags inside the framework, and the weight of the whole construction was 9 tons--a ton less than that of the first Zeppelin airship. Three vertical planes at front and rear controlled horizontal steering, while rise and fall was controlled by horizontal planes arranged in box form. Accident attended the first trial of this second airship, which took place over the Bodensee on November 30th, 1905, 'It had been intended to tow the raft, to which it was anchored, further from the shore against the wind. But the water was too low to allow the use of the raft. The balloon was therefore mounted on pontoons, pulled out into the lake, and taken in tow by a motor-boat. It was caught by a strong wind which was blowing from the shore, and driven ahead at such a rate that it overtook the motor-boat. The tow rope was therefore at once cut, but it unexpectedly formed into knots and became entangled with the airship, pulling the front end down into the water. The balloon was then caught by the wind and lifted into the air, when the propellers were set in motion. The front end was at this instant pointing in a downward direction, and consequently it shot into the water, where it was found necessary to open the valves.'[14]

The damage done was repaired within six weeks, and the second trial was made on January 17th, 1906. The lifting force was too great for the weight, and the dirigible jumped immediately to 1,500 feet. The propellers were started, and the dirigible brought to a lower level, when it was found possible to drive against the wind. The steering arrangements were found too sensitive, and the motors were stopped, when the vessel was carried by the wind until it was over land--it had been intended that the trial should be completed over water. A descent was successfully accomplished and the dirigible was anchored for the night, but a gale caused it so much damage that it had to be broken up. It had achieved a speed of 30 feet per second with the motors developing only 36 horse-power and, gathering from this what speed might have been accomplished with the full 170 horse-power, Zeppelin set about the construction of No. 3, with which a number of successful voyages were made, proving the value of the type for military purposes.

No. 4 was the most notable of the early Zeppelins, as much on account of its disastrous end as by reason of any superior merit in comparison with No. 3. The main innovation consisted in attaching a triangular keel to the under side of the envelope, with two gaps beneath which the cars were suspended. Two Daimler Mercedes motors of 110 horse-power each were placed one in each car, and the vessel carried sufficient fuel for a 60-hour cruise with the motors running at full speed. Each motor drove a pair of three-bladed metal propellers rigidly attached to the framework of the envelope and about 15 feet in diameter. There was a vertical rudder at the stern of the envelope and horizontal controlling planes were fixed on the sides of the envelope. The best performances and the end of this dirigible were summarised as follows by Major Squier:--

'Its best performances were two long trips performed during the summer of 1908. The first, on July 4th, lasted exactly 12 hours, during which time it covered a distance of 235 miles, crossing the mountains to Lucerne and Zurich, and returning to the balloon-house near Friedrichshafen, on Lake Constance. The average speed on this trip was 32 miles per hour. On August 4th, this airship attempted a 24-hour

---
14 Hildebrandt, Airships Past and Present.

flight, which was one of the requirements made for its acceptance by the Government. It left Friedrichshafen in the morning with the intention of following the Rhine as far as Mainz, and then returning to its starting-point, straight across the country. A stop of 3 hours 30 minutes was made in the afternoon of the first day on the Rhine, to repair the engine. On the return, a second stop was found necessary near Stuttgart, due to difficulties with the motors, and some loss of gas. While anchored to the ground, a storm arose which broke loose the anchorage, and, as the balloon rose in the air, it exploded and took fire (due to causes which have never been actually determined and published) and fell to the ground, where it was completely destroyed. On this journey, which lasted in all 31 hours 15 minutes, the airship was in the air 20 hours 45 minutes, and covered a total distance of 378 miles.

'The patriotism of the German nation was aroused. Subscriptions were immediately started, and in a short space of time a quarter of a million pounds had been raised. A Zeppelin Society was formed to direct the expenditure of this fund. Seventeen thousand pounds has been expended in purchasing land near Friedrichshafen; workshops were erected, and it was announced that within one year the construction of eight airships of the Zeppelin type would be completed. Since the disaster to 'Zeppelin IV.' the Crown Prince of Germany made a trip in 'Zeppelin No. 3,' which had been called back into service, and within a very few days the German Emperor visited Friedrichshafen for the purpose of seeing the airship in flight. He decorated Count Zeppelin with the order of the Black Eagle. German patriotism and enthusiasm has gone further, and the "German Association for an Aerial Fleet" has been organised in sections throughout the country. It announces its intention of building 50 garages (hangars) for housing airships.'

By January of 1909, with well over a quarter of a million in hand for the construction of Zeppelin airships, No. 3 was again brought out, probably in order to maintain public enthusiasm in respect of the possible new engine of war. In March of that year No. 3 made a voyage which lasted for 4 hours over and in the vicinity of Lake Constance; it carried 26 passengers for a distance of nearly 150 miles.

Before the end of March, Count Zeppelin determined to voyage from Friedrichshafen to Munich, together with the crew of the airship and four military officers. Starting at four in the morning and ascertaining their route from the lights of railway stations and the ringing of bells in the towns passed over, the journey was completed by nine o'clock, but a strong south-west gale prevented the intended landing. The airship was driven before the wind until three o'clock in the afternoon, when it landed safely near Dingolfing; by the next morning the wind had fallen considerably and the airship returned to Munich and landed on the parade ground as originally intended. At about 3.30 in the afternoon, the homeward journey was begun, Friedrichshafen being reached at about 7.30.

These trials demonstrated that sufficient progress had been made to justify the construction of Zeppelin airships for use with the German army. No. 3 had been manoeuvred safely if not successfully in half a gale of wind, and henceforth it was known as 'SMS. Zeppelin I.,' at the bidding of the German Emperor, while the construction of 'SMS. Zeppelin II.' was rapidly proceeded with. The fifth construction of Count Zeppelin's was 446 feet in length, 42 1/2 feet in diameter, and contained 530,000 cubic feet of hydrogen gas in 17 separate compartments. Trial flights were made on the 26th May, 1909, and a week later she made a record voyage of 940 miles, the route being from Lake Constance over Ulm, Nuremberg, Leipzig, Bitterfeld, Weimar, Heilbronn, and Stuttgart, descending near Goppingen; the time occupied in the flight was upwards of 38 hours.

In landing, the airship collided with a pear-tree, which damaged the bows and tore open two sections of the envelope, but repairs on the spot enabled the return journey to Friedrichshafen to be begun 24 hours later. In spite of the mishap the Zeppelin had once more proved itself as a possible engine of war, and thenceforth Germany pinned its faith to the dirigible, only developing the aeroplane to such an extent as to keep abreast of other nations. By the outbreak of war, nearly 30 Zeppelins had been constructed; considerably more than half of these were destroyed in various ways, but the experiments carried on with each example of the type permitted of improvements being made. The first fatality occurred in September, 1913, when the fourteenth Zeppelin to be constructed, known as Naval Zeppelin L.1, was wrecked in the North Sea by a sudden storm and her crew of thirteen were drowned. About three weeks after this, Naval Zeppelin L.2, the eighteenth in order of building, exploded in mid-air while manoeuvring over Johannisthal. She was carrying a crew of 25, who were all killed.

By 1912 the success of the Zeppelin type brought imitators. Chief among them was the Schutte-Lanz, a Mannheim firm, which produced a rigid dirigible with a wooden framework, wire braced. This was not a cylinder like the Zeppelin, but reverted to the cigar shape and contained about the same amount of gas as the Zeppelin type. The Schutte-Lanz was made with two gondolas rigidly attached to the envelope in which the gas bags were placed. The method of construction involved greater weight than was the case with the Zeppelin, but the second of these vessels, built with three gondolas containing engines, and a navigating cabin built into the hull of the airship itself, proved quite successful as a naval scout until wrecked on the islands off the coast of Denmark late in 1914. The last Schutte-Lanz to be constructed was used by the Germans for raiding England, and was eventually brought down in flames at Cowley.

V. BRITISH AIRSHIP DESIGN

As was the case with the aeroplane, Great Britain left France and Germany to make the running in the early days of airship construction; the balloon section of the Royal Engineers was compelled to confine its energies to work with balloons pure and simple until well after the twentieth century had dawned, and such experiments as were made in England were done by private initiative. As far back as 1900 Doctor Barton built an airship at the Alexandra Palace and voyaged across London in it. Four years later Mr E. T. Willows of Cardiff produced the first successful British dirigible, a semi-rigid 74 feet in length and 18 feet in diameter, engined with a 7 horse-power Peugot twin-cylindered motor. This drove a two-bladed propeller at the stern for propulsion, and also actuated a pair of auxiliary propellers at the front which could be varied in their direction so as to control the right and left movements of the airship. This device was patented and the patent was taken over by the British Government, which by 1908 found Mr Willow's work of sufficient interest to regard it as furnishing data for experiment at the balloon factory at Farnborough. In 1909, Willows steered one of his dirigibles to London from Cardiff in a little less than ten hours, making an average speed of over 14 miles an hour. The best speed accomplished was probably considerably greater than this, for at intervals of a few miles, Willows descended near the earth to ascertain his whereabouts with the help of a megaphone. It must be added that he carried a compass in addition to his megaphone. He set out for Paris in November of 1910, reached the French coast, and landed near Douai. Some damage was sustained in this landing, but, after repair, the trip to Paris was completed.

Meanwhile the Government balloon factory at Farnborough began airship construction in 1907; Colonel Capper, R.E., and S. F. Cody were jointly concerned in the production of a semi-rigid. Fifteen thicknesses of goldbeaters' skin--about the most expensive covering obtainable--were used for the envelope, which was 25 feet in diameter. A slight shower of rain in which the airship was caught led to its wreckage, owing to the absorbent quality of the goldbeaters' skin, whereupon Capper and Cody set to work to reproduce the airship and its defects on a larger scale. The first had been named 'Nulli Secundus' and the second was named 'Nulli Secundus II.' Punch very appropriately suggested that the first vessel ought to have been named 'Nulli Primus,' while a possible third should be christened 'Nulli Tertius.' 'Nulli Secundus II.' was fitted with a 100 horse-power engine and had an envelope of 42 feet in diameter, the goldbeaters' skin being covered in fabric and the car being suspended by four bands which encircled the balloon envelope. In October of 1907, 'Nulli Secundus II.' made a trial flight from Farnborough to London and was anchored at the Crystal Palace. The wind sprung up and took the vessel away from its mooring ropes, wrecking it after the one flight.

Stagnation followed until early in 1909, when a small airship fitted with two 12 horse-power motors and named the 'Baby' was turned out from the balloon factory. This was almost egg-shaped, the blunt end being forward, and three inflated fins being placed at the tail as control members. A long car with rudder and elevator at its rear-end carried the engines and crew; the 'Baby' made some fairly successful flights and gave a good deal of useful data for the construction of later vessels.

Next to this was 'Army Airship 2A 'launched early in 1910 and larger, longer, and narrower in design than the Baby. The engine was an 80 horse-power Green motor which drove two pairs of propellers; small inflated control members were fitted at the stern end of the envelope, which was 154 feet in length. The suspended car was 84 feet long, carrying both engines and crew, and the Willows idea of swivelling propellers for governing the direction was used in this vessel. In June of that year a new, small-type dirigible, the 'Beta,' was produced, driven by a 30 horse-power Green engine with which she flew over 3,000 miles. She was the most successful British dirigible constructed up to that time, and her successor, the 'Gamma,' was built on similar lines. The 'Gamma' was a larger vessel, however, produced in 1912, with flat, controlling fins and rudder at the rear end of the envelope, and with the conventional long car suspended at some distance beneath the gas bag. By this time, the mooring mast, carrying a cap of which the concave side fitted over the convex nose of the airship, had been originated. The cap was swivelled, and, when attached to it, an airship was held nose on to the wind, thus reducing by more than half the dangers attendant on mooring dirigibles in the open.

Private subscription under the auspices of the Morning Post got together sufficient funds in 1910 for the purchase of a Lebaudy airship, which was built in France, flown across the Channel, and presented to the Army Airship Fleet. This dirigible was 337 feet long, and was driven by two 135 horse-power Panhard motors, each of which actuated two propellers. The journey from Moisson to Aldershot was completed at a speed of 36 miles an hour, but the airship was damaged while being towed into its shed. On May of the following year, the Lebaudy was brought out for a flight, but, in landing, the guide rope fouled in trees and sheds and brought the airship broadside on to the wind; she was driven into some trees and wrecked to such an exteent that rebuilding was considered an impossibility. A Clement Bayard, bought by the army airship section, became scrap after even less flying than had been accomplished by the Lebaudy.

In April of 1910, the Admiralty determined on a naval air service, and set about the production of rigid airships which should be able to compete with Zeppelins as naval scouts. The construction was entrusted to Vickers, Ltd., who set about the task at their Barrow works and built something which, when tested after a year's work, was found incapable of lifting its own weight. This defect was remedied by a series of alterations, and meanwhile the unofficial title of 'Mayfly' was given to the vessel.

Taken over by the Admiralty before she had passed any flying tests, the 'Mayfly' was brought out on September 24th, 1911, for a trial trip, being towed out from her shed by a tug. When half out from the shed, the envelope was caught by a light cross-wind, and, in spite of the pull from the tug, the great fabric broke in half, nearly drowning the crew, who had to dive in order to get clear of the wreckage.

There was considerable similarity in form, though not in performance, between the Mayfly and the prewar Zeppelin. The former was 510 feet in length, cylindrical in form, with a diameter of 48 feet, and divided into 19 gas-bag compartments. The motive power consisted of two 200 horse-power Wolseley engines. After its failure, the Naval Air Service bought an Astra-Torres airship from France and a Parseval from Germany, both of which proved very useful in the early days of the War, doing patrol work over the Channel before the Blimps came into being.

Early in 1915 the 'Blimp' or 'S.S.' type of coastal airship was evolved in response to the demand for a vessel which could be turned out quickly and in quantities. There was urgent demand, voiced by Lord Fisher, for a type of vessel capable of maintaining anti-submarine patrol off the British coasts, and the first S.S. airships were made by combining a gasbag with the most available type of aeroplane fuselage and engine, and fitting steering gear. The 'Blimp' consisted of a B.E. fuselage with engine and geared-down propeller, and seating for pilot and observer, attached to an envelope about 150 feet in length. With a speed of between 35 and 40 miles an hour, the 'Blimp' had a cruising capacity of about ten hours; it was fitted with wireless set, camera, machine-gun, and bombs, and for submarine spotting and patrol work generally it proved invaluable, though owing to low engine power and comparatively small size, its uses were restricted to reasonably fair weather. For work farther out at sea and in all weathers, airships known as the coast patrol type, and more commonly as 'coastals,' were built, and later the 'N.S.' or North Sea type, still larger and more weather-worthy, followed. By the time the last year of the War came, Britain led the world in the design of non-rigid and semi-rigid dirigibles. The 'S.S.' or 'Blimp' had been improved to a speed of 50 miles an hour, carrying a crew of three, and the endurance record for the type was 18 1/2 hours, while one of them had reached a height of 10,000 feet. The North Sea type of non-rigid was capable of travelling over 20 hours at full speed, or forty hours at cruising speed, and the number of non-rigids belonging to the British Navy exceeded that of any other country.

It was owing to the incapacity--apparent or real--of the British military or naval designers to produce a satisfactory rigid airship that the 'N.S.' airship was evolved. The first of this type was produced in 1916, and on her trials she was voted an unqualified success, in consequence of which the building of several more was pushed on. The envelope, of 360,000 cubic feet capacity, was made on the Astra-

Torres principle of three lobes, giving a trefoil section. The ship carried four fins, to three of which the elevator and rudder flaps were attached; petrol tanks were placed inside the envelope, under which was rigged a long covered-in car, built up of a light steel tubular framework 35 feet in length. The forward portion was covered with duralumin sheeting, an aluminium alloy which, unlike aluminium itself, is not affected by the action of sea air and water, and the remainder with fabric laced to the framework. Windows and port-holes were provided to give light to the crew, and the controls and navigating instruments were placed forward, with the sleeping accommodation aft. The engines were mounted in a power unit structure, separate from the car and connected by wooden gang ways supported by wire cables. A complete electrical installation of two dynamos and batteries for lights, signalling lamps, wireless, telephones, etc., was carried, and the motive power consisted of either two 250 horse-power Rolls-Royce engines or two 240 horse-power Fiat engines. The principal dimensions of this type are length 262 feet, horizontal diameter 56 feet 9 inches, vertical diameter 69 feet 3 inches. The gross lift is 24,300 lbs. and the disposable lift without crew, petrol, oil, and ballast 8,500 lbs. The normal crew carried for patrol work was ten officers and men. This type holds the record of 101 hours continuous flight on patrol duty.

In the matter of rigid design it was not until 1913 that the British Admiralty got over the fact that the 'Mayfly' would not, and decided on a further attempt at the construction of a rigid dirigible. The contract for this was signed in March of 1914; work was suspended in the following February and begun again in July, 1915, but it was not until January of 1917 that the ship was finished, while her trials were not completed until March of 1917, when she was taken over by the Admiralty. The details of the construction and trial of this vessel, known as 'No. 9,' go to show that she did not quite fill the contract requirements in respect of disposable lift until a number of alterations had been made. The contract specified that a speed of at least 45 miles per hour was to be attained at full engine power, while a minimum disposable lift of 5 tons was to be available for movable weights, and the airship was to be capable of rising to a height of 2,000 feet. Driven by four Wolseley Maybach engines of 180 horse-power each, the lift of the vessel was not sufficient, so it was decided to remove the two engines in the after car and replace them by a single engine of 250 horsepower. With this the vessel reached the contract speed of 45 miles per hour with a cruising radius of 18 hours, equivalent to 800 miles when the engines were running at full speed. The vessel served admirably as a training airship, for, by the time she was completed, the No. 23 class of rigid airship had come to being, and thus No. 9 was already out of date.

Three of the 23 class were completed by the end of 1917; it was stipulated that they should be built with a speed of at least 55 miles per hour, a minimum disposable lift of 8 tons, and a capability of rising at an average rate of not less than 1,000 feet per minute to a height of 3,000 feet. The motive power consisted of four 250 horse-power Rolls-Royce engines, one in each of the forward and after cars and two in a centre car. Four-bladed propellers were used throughout the ship.

A 23X type followed on the 23 class, but by the time two ships had been completed, this was practically obsolete. The No. 31 class followed the 23X; it was built on Schutte-Lanz lines, 615 feet in length, 66 feet diameter, and a million and a half cubic feet capacity. The hull was similar to the later types of Zeppelin in shape, with a tapering stern and a bluff, rounded bow. Five cars each carrying a 250 horse-power Rolls-Royce engine, driving a single fixed propeller, were fitted, and on her trials R.31 performed well, especially in the matter of speed. But the experiment of constructing in wood in the Schutte-Lanz way adopted with this vessel resulted in failure eventually, and the type was abandoned.

Meanwhile, Germany had been pushing forward Zeppelin design and straining every nerve in the improvement of rigid dirigible construction, until L.33 was evolved; she was generally known as a super-Zeppelin, and on September 24th, 1916, six weeks after her launching, she was damaged by gun-fire in a raid over London, being eventually compelled to come to earth at Little Wigborough in Essex. The crew gave themselves up after having set fire to the ship, and though the fabric was totally destroyed, the structure of the hull remained intact, so that just as Germany was able to evolve the Gotha bomber from the Handley-Page delivered at Lille, British naval constructors were able to evolve the R.33 type of airship from the Zeppelin framework delivered at Little Wigborough. Two vessels, R.33 and R.34, were laid down for completion; three others were also put down for construction, but, while R.33 and R.34 were built almost entirely from the data gathered from the wrecked L.33, the three later vessels embody more modern design, including a number of improvements, and more especially greater disposable lift. It has been commented that while the British authorities were building R.33 and R.34, Germany constructed 30 Zeppelins on 4 slips, for which reason it may be reckoned a matter for congratulation that the rigid airship did not decide the fate of the War. The following particulars of construction of the R.33 and R.34 types are as given by Major Whale in his survey of British Airships:--

'In all its main features the hull structure of R.33 and R.34 follows the design of the wrecked German Zeppelin airship L.33. 'The hull follows more nearly a true stream-line shape than in the previous ships constructed of duralumin, in which a greater proportion of the greater length was parallel-sided. The Germans adopted this new shape from the Schutte-Lanz design and have not departed from this practice. This consists of a short, parallel body with a long, rounded bow and a long tapering stem culminating in a point. The overall length of the ship is 643 feet with a diameter of 79 feet and an extreme height of 92 feet.

'The type of girders in this class has been much altered from those in previous ships. The hull is fitted with an internal triangular keel throughout practically the entire length. This forms the main corridor of the ship, and is fitted with a footway down the centre for its entire length. It contains water ballast and petrol tanks, bomb storage and crew accommodation, and the various control wires, petrol pipes, and electric leads are carried along the lower part.

'Throughout this internal corridor runs a bridge girder, from which the petrol and water ballast tanks are supported. These tanks are so arranged that they can be dropped clear of the ship. Amidships is the cabin space with sufficient room for a crew of twenty-five. Hammocks can be swung from the bridge girder before mentioned.

'In accordance with the latest Zeppelin practice, monoplane rudders and elevators are fitted to the horizontal and vertical fins.

'The ship is supported in the air by nineteen gas bags, which give a total capacity of approximately two million cubic feet of gas. The gross lift works out at approximately 59 1/2 tons, of which the total fixed weight is 33 tons, giving a disposable lift of 26 1/2 tons.

'The arrangement of cars is as follows: At the forward end the control car is slung, which contains all navigating instruments and the various controls. Adjoining this is the wireless cabin, which is also fitted for wireless telephony. Immediately aft of this is the forward power car containing one engine, which gives the appearance that the whole is one large car.

'Amidships are two wing cars, each containing a single engine. These are small and just accommodate the engines with sufficient room for mechanics to attend to them. Further aft is another larger car which contains an auxiliary control position and two engines.

'It will thus be seen that five engines are installed in the ship; these are all of the same type and horsepower, namely, 250 horse-power Sunbeam. R.33 was constructed by Messrs Armstrong, Whitworth, Ltd.; while her sister ship R.34 was built by Messrs Beardmore on the Clyde.'

Of the two vessels, R.34 appeared rather more airworthy than her sister ship; the lift of the ship justified the carrying of a greater quantity of fuel than had been provided for, and, as she was considered suitable for making a Transatlantic crossing, extra petrol tanks were fitted in the hull and a new type of outer cover was fitted with a view to her making the Atlantic crossing. She made a 21-hour cruise over the North of England and the South of Scotland at the end of May, 1919, and subsequently went for a longer cruise over Denmark, the Baltic, and the north coast of Germany, remaining in the air for 56 hours in spite of very bad weather conditions. Finally, July 2nd was selected as the starting date for the cross Atlantic flight; the vessel was commanded by Major G. H. Scott, A.F.C., with Captain G. S. Greenland as first officer, Second-Lieut. H. F. Luck as second officer, and Lieut. J. D. Shotter as engineer officer. There were also on board Brig.-Gen. E. P. Maitland, representing the Air Ministry, Major J. E. M. Pritchard, representing the Admiralty, and Lieut.-Col. W. H. Hemsley of the Army Aviation Department. In addition to eight tons of petrol, R.34 carried a total number of 30 persons from East Fortune to Long Island, N.Y.

There being no shed in America capable of accommodating the airship, she had to be moored in the open for refilling with fuel and gas, and to make the return journey almost immediately.

Brig.-Gen. Maitland's account of the flight, in itself a record as interesting as valuable, divides the outward journey into two main stages, the first from East Fortune to Trinity Bay, Newfoundland, a distance of 2,050 sea miles, and the second and more difficult stage to Mineola Field, Long Island, 1,080 sea miles. An easy journey was experienced until Newfoundland was reached, but then storms and electrical disturbances rendered it necessary to alter the course, in consequence of which petrol began to run short. Head winds rendered the shortage still more acute, and on Saturday, July 5th, a wireless signal was sent out asking for destroyers to stand by to tow. However, after an anxious night, R.33 landed safely at Mineola Field at 9.55 a.m. on July 6th, having accomplished the journey in 108 hours 12 minutes.

She remained at Mineola until midnight of July 9th, when, although it had been intended that a start should be made by daylight for the benefit of New York spectators, an approaching storm caused preparations to be advanced for immediate departure. She set out at 5.57 a.m. by British summer time, and flew over New York in the full glare of hundreds of searchlights before heading out over the Atlantic. A following wind assisted the return voyage, and on July 13th, at 7.57 a.m., R.34 anchored at Pulham, Norfolk, having made the return journey in 75 hours 3 minutes, and proved the suitability of the dirigible for Transatlantic commercial work. R.80, launched on July 19th, 1920, afforded further proof, if this were needed.

It is to be noted that nearly all the disasters to airships have been caused by launching and landing--the type is safe enough in the air, under its own power, but its bulk renders it unwieldy for ground handling. The German system of handling Zeppelins in and out of their sheds is, so far, the best devised: this consists of heavy trucks running on rails through the sheds and out at either end; on descending, the trucks are run out, and the airship is securely attached to them outside the shed; the trucks are then run back into the shed, taking the airship with them, and preventing any possibility of the wind driving the envelope against the side of the shed before it is safely housed; the reverse process is adopted in launching, which is thus rendered as simple as it is safe.

VI. THE AIRSHIP COMMERCIALLY

Prior to the war period, between the years 1910 and 1914, a German undertaking called the Deutsche Luftfahrt Actien Gesellschaft conducted a commercial Zeppelin service in which four airships known as the Sachsan, Hansa, Victoria Louise, and Schwaben were used. During the four years of its work, the company carried over 17,000 passengers, and over 100,000 miles were flown without incurring one fatality and with only minor and unavoidable accidents to the vessels composing the service. Although a number of English notabilities made voyages in these airships, the success of this only experiment in commercial aerostation seems to have been forgotten since the war. There was beyond doubt a military aim in this apparently peaceful use of Zeppelin airships; it is past question now that all Germany's mechanical development in respect of land sea, and air transport in the years immediately preceding the war, was accomplished with the ulterior aim of military conquest, but, at the same time, the running of this service afforded proof of the possibility of establishing a dirigible service for peaceful ends, and afforded proof too, of the value of the dirigible as a vessel of purely commercial utility.

In considering the possibility of a commercial dirigible service, it is necessary always to bear in mind the disadvantages of first cost and upkeep as compared with the aeroplane. The building of a modern rigid is an exceedingly costly undertaking, and the provision of an efficient supply of hydrogen gas to keep its compartments filled is a very large item in upkeep of which the heavier-than-air machine goes free. Yet the future of commercial aeronautics so far would seem to lie with the dirigible where very long voyages are in question. No matter how the aeroplane may be improved, the possibility of engine failure always remains as a danger for work over water. In seaplane or flying boat form, the danger is still present in a rough sea, though in the American Transatlantic flight, N.C.3, taxi-ing 300 miles to the Azores after having fallen to the water, proved that this danger is not so acute as is generally assumed. Yet the multiple-engined rigid, as R.34 showed on her return voyage, may have part of her power plant put out of action altogether and still complete her voyage very successfully, which, in the case of mail carrying and services run strictly to time, gives her an enormous advantage over the heavier-than-air machine.

'For commercial purposes,' General Sykes has remarked, 'the airship is eminently adapted for long distance journeys involving non-stop flights. It has this inherent advantage over the aeroplane, that while there appears to be a limit to the range of the aeroplane as at present constructed, there is practically no limit whatever to that of the airship, as this can be overcome by merely increasing the size. It thus appears that for such journeys as crossing the Atlantic, or crossing the Pacific from the west coast of America to Australia or Japan, the airship will be peculiarly suitable. It having been conceded that the scope of the airship is long distance travel, the only type which need be considered for this purpose is the rigid. The rigid airship is still in an embryonic state, but sufficient has already been accomplished in this country, and more particularly in Germany, to show that with increased capacity there is no reason why, within a few years' time, airships should not be built capable of completing the circuit of the globe and of conveying sufficient passengers and merchandise to render such an undertaking a paying proposition.'

The British R.38 class, embodying the latest improvements in airship design outside Germany, gives a gross lift per airship of 85 tons and a net lift of about 45 tons. The capacity of the gas bags is about two and three-quarter million cubic feet, and, travelling at the rate of 45 miles per hour, the cruising range of the vessel is estimated at 8.8 days. Six engines, each of 350 horse-power, admit of an extreme speed of 70 miles per hour if necessary.

The last word in German design is exemplified in the rigids L.70 and L.71, together with the commercial airship 'Bodensee.' Previous to the construction of these, the L.65 type is noteworthy as being the first Zeppelin in which direct drive of the propeller was introduced, together with an improved and lighter type of car. L.70 built in 1918 and destroyed by the British naval forces, had a speed of about 75 miles per hour; L.71 had a maximum speed of 72 miles per hour, a gas bag capacity of 2,420,000 cubic feet, and a length of 743 feet, while the total lift was 73 tons. Progress in design is best shown by the progress in useful load; in the L.70 and L.71 class, this has been increased to 58.3 per cent, while in the Bodensee it was ever higher.

As was shown in R.34's American flight, the main problem in connection with the commercial use of dirigibles is that of mooring in the open. The nearest to a solution of this problem, so far, consists in the mast carrying a swivelling cap; this has been tried in the British service with a non-rigid airship, which was attached to a mast in open country in a gale of 52 miles an hour without the slightest damage to the airship. In its commercial form, the mast would probably take the form of a tower, at the top of which the cap would revolve so that the airship should always face the wind, the tower being used for embarkation and disembarkation of passengers and the provision of fuel and gas. Such a system would render sheds unnecessary except in case of repairs, and would enormously decrease the establishment charges of any commercial airship.

All this, however, is hypothetical. Remains the airship of to-day, developed far beyond the promise of five years ago, capable, as has been proved by its achievements both in Britain and in Germany, of undertaking practically any given voyage with success.

VII. KITE BALLOONS

As far back as the period of the Napoleonic wars, the balloon was given a place in warfare, but up to the Franco-Prussian Prussian War of 1870-71 its use was intermittent. The Federal forces made use of balloons to a small extent in the American Civil War; they came to great prominence in the siege of Paris, carrying out upwards of three million letters and sundry carrier pigeons which took back messages into the besieged city. Meanwhile, as captive balloons, the German and other armies used them for observation and the direction of artillery fire. In this work the ordinary spherical balloon was at a grave disadvantage; if a gust of wind struck it, the balloon was blown downward and down wind, generally twirling in the air and upsetting any calculations and estimates that might be made by the observers, while in a wind of 25 miles an hour it could not rise at all. The rotatory movement caused by wind was stopped by an experimenter in the Russo-Japanese war, who fixed to the captive observation balloons a fin which acted as a rudder. This did not stop the balloon from being blown downward and away from its mooring station, but this tendency was overcome by a modification designed in Germany by the Parseval-Siegsfield Company, which originated what has since become familiar as the 'Sausage' or kite balloon. This is so arranged that the forward end is tilted up into the wind, and the underside of the gas bag, acting as a plane, gives the balloon a lifting tendency in a wind, thus counteracting the tendency of the wind to blow it downward and away from its mooring station. Smaller bags are fitted at the lower and rear end of the balloon with openings that face into the wind; these are thus kept inflated, and they serve the purpose of a rudder, keeping the kite balloon steady in the air.

Various types of kite balloon have been introduced; the original German Parseval-Siegsfield had a single air bag at the stern end, which was modified to two, three, or more lobes in later varieties, while an American experimental design attempted to do away with the attached lobes altogether by stringing out a series of small air bags, kite fashion, in rear of the main envelope. At the beginning of the War, Germany alone had kite balloons, for the authorities of the Allied armies considered that the bulk of such a vessel rendered it too conspicuous a mark to permit of its being serviceable. The Belgian arm alone possessed two which, on being put into service, were found extremely useful. The French followed by constructing kite balloons at Chalais Meudon, and then, after some months of hostilities and with the example of the Royal Naval Air Service to encourage them, the British military authorities finally took up the construction and use of kite balloons for artillery-spotting and general observation purposes. Although many were brought down by gun-fire, their uses far outweighed their disadvantages, and toward the end of the War, hardly a mile of front was without its 'Sausage.'

For naval work, kite balloons were carried in a specially constructed hold in the forepart of certain vessels; when required for use, the covering of the hold was removed, the kite balloon inflated and released to the required height by means of winches as in the case of the land work. The perfecting of the 'Coastal' and N.S. types of airship, together with the extension of wireless telephony between airship and cruiser or other warship, in all probability will render the use of the kite balloon unnecessary in connection with naval scouting. But, during the War, neither wireless telephony nor naval airships had developed sufficiently to render the Navy independent of any means that might come to hand, and the fitting of kite balloons in this fashion filled a need of the times.

A necessary accessory of the kite balloon is the parachute, which has a long history. Da Vinci and Veranzio appear to have been the first exponents, the first in the theory and the latter in the practice of parachuting. Montgolfier experimented at Annonay before he constructed his first hot air-balloon, and in 1783 a certain Lenormand dropped from a tree in a parachute. Blanchard the balloonist made a spectacle of parachuting, and made it a financial success; Cocking, in 1836, attempted to use an inverted form of parachute; taken up to a height of 3,000 feet, he was cut adrift, when the framework of the parachute collapsed and Cocking was killed.

The rate of fall is slow in parachuting to the ground. Frau Poitevin, making a descent from a height of 6,000 feet, took 45 minutes to reach the ground, and, when she alighted, her husband, who had taken her up, had nearly got his balloon packed up. Robertson, another parachutist is said to have descended from a height of 10,000 feet in 35 minutes, or at a rate of nearly 5 feet per second. During the War Brigadier-General Maitland made a parachute descent from a height of 10,000 feet, the time taken being about 20 minutes.

The parachute was developed considerably during the War period, the main requirement, that of certainty in opening, being considerably developed. Considered a necessary accessory for kite balloons, the parachute was also partially adopted for use with aeroplanes in the later War period, when it was contended that if a machine were shot down in flames, its occupants would be given a far better chance of escape if they had parachutes. Various trials were made to demonstrate the extreme efficiency of the parachute in modern form, one of them being a descent from the upper ways of the Tower Bridge to the waters of the Thames, in which short distance the 'Guardian Angel' type of parachute opened and cushioned the descent for its user.

For dirigibles, balloons, and kite balloons the parachute is an essential. It would seem to be equally essential in the case of heavier-than-air machines, but this point is still debated. Certainly it affords the occupant of a falling aeroplane a chance, no matter how slender, of reaching the ground in safety, and, for that reason, it would seem to have a place in aviation as well as in aerostation.

PART IV. ENGINE DEVELOPMENT
I. THE VERTICAL TYPE

The balloon was but a year old when the brothers Robert, in 1784 attempted propulsion of an aerial vehicle by hand-power, and succeeded, to a certain extent, since they were able to make progress when there was only a slight wind to counteract their work. But, as may be easily understood, the manual power provided gave but a very slow speed, and in any wind it all the would-be airship became an uncontrolled balloon.

Henson and Stringfellow, with their light steam engines, were first to attempt conquest of the problem of mechanical propulsion in the air; their work in this direction is so fully linked up with their constructed models that it has been outlined in the section dealing with the development of the aeroplane. But, very shortly after these two began, there came into the field a Monsieur Henri Giffard, who first achieved success in the propulsion by mechanical means of dirigible balloons, for his was the first airship to fly against the wind. He employed a small steam-engine developing about 3 horse-power and weighing 350 lbs. with boiler, fitting the whole in a car suspended from the gas-bag of his dirigible. The propeller which this engine worked was 11 feet in diameter, and the inventor, who made several flights, obtained a speed of 6 miles an hour against a slight wind. The power was not sufficient to render the invention practicable, as the dirigible could only be used in calm weather, but Giffard was sufficiently encouraged by his results to get out plans for immense dirigibles, which through lack of funds he was unable to construct. When, later, his invention of the steam-injector gave him the means he desired, he became blind, and in 1882 died, having built but the one famous dirigible.

This appears to have been the only instance of a steam engine being fitted to a dirigible; the inherent disadvantage of this form of motive power is that a boiler to generate the steam must be carried, and this, together with the weight of water and fuel, renders the steam engine uneconomical in relation to the lift either of plane or gas-bag. Again, even if the weight could be brought down to a reasonable amount, the attention required by steam plant renders it undesirable as a motive power for aircraft when compared with the internal combustion engine.

Maxim, in Artificial and Natural Flight, details the engine which he constructed for use with his giant experimental flying machine, and his description is worthy of reproduction since it is that of the only steam engine besides Giffard's, and apart from those used for the propulsion of models, designed for driving an aeroplane. 'In 1889,' Maxim says, 'I had my attention drawn to some very thin, strong, and comparatively cheap tubes which were being made in France, and it was only after I had seen these tubes that I seriously considered the question of making a flying machine. I obtained a large quantity of them and found that they were very light, that they would stand enormously high pressures, and generate a very large quantity of steam. Upon going into a mathematical calculation of the whole subject, I found that it would be possible to make a machine on the aeroplane system, driven by a steam engine, which would be sufficiently strong to lift itself into the air. I first made drawings of a steam engine, and a pair of these engines was afterwards made. These engines are constructed, for the most part, of a very high grade of cast steel, the cylinders being only 3/32 of an inch thick, the crank shafts hollow, and every part as strong and light as possible. They are compound, each having a high-pressure piston with an area of 20 square inches, a low-pressure piston of 50.26 square inches, and a common stroke of 1 foot. When first finished they were found to weigh 300 lbs. each; but after putting on the oil cups, felting, painting, and making some slight alterations, the weight was brought up to 320 lbs. each, or a total of 640 lbs. for the two engines, which have since developed 362 horsepower with a steam pressure of 320 lbs. per square inch.'

The result is remarkable, being less than 2 lbs. weight per horse-power, especially when one considers the state of development to which the steam engine had attained at the time these experiments were made. The fining down of the internal combustion engine, which has done so much to solve the problems of power in relation to weight for use with aircraft, had not then been begun, and Maxim had nothing to guide him, so far as work on the part of his predecessors was concerned, save the experimental engines of Stringfellow, which, being constructed on so small a scale in comparison with his own, afforded little guidance. Concerning the factor of power, he says: 'When first designing this engine, I did not know how much power I might require from it. I thought that in some cases it might be necessary to allow the high-pressure steam to enter the low-pressure cylinder direct, but as this would involve a considerable loss, I constructed a species of injector. This injector may be so adjusted that when the steam in the boiler rises above a certain predetermined point, say 300 lbs., to the square inch, it opens a valve and escapes past the high-pressure cylinder instead of blowing off at the safety valve. In escaping through this valve, a fall of about 200 lbs. pressure per square inch is made to do work on the surrounding steam and drive it forward in the pipe, producing a pressure on the low-pressure piston considerably higher than the back-pressure on the high-pressure piston. In this way a portion of the work which would otherwise be lost is utilised, and it is possible, with an unlimited supply of steam, to cause the engines to develop an enormous amount of power.'

With regard to boilers, Maxim writes,

'The first boiler which I made was constructed something on the Herreshof principle, but instead of having one simple pipe in one very long coil, I used a series of very small and light pipes, connected in such a manner that there was a rapid circulation through the whole--the tubes increasing in size and number as the steam was generated. I intended that there should be a pressure of about 100 lbs. more on the feed water end of the series than on the steam end, and I believed that this difference in pressure would be sufficient to ensure direct and positive circulation through every tube in the series. The first boiler was exceedingly light, but the workmanship, as far as putting the tubes together was concerned, was very bad, and it was found impossible to so adjust the supply of water as to make dry steam without overheating and destroying the tubes.

'Before making another boiler I obtained a quantity of copper tubes, about 8 feet long, 3/8 inch external diameter, and 1/50 of an inch thick. I subjected about 100 of these tubes to an internal pressure of 1 ton per square inch of cold kerosene oil, and as none of them leaked I did not test any more, but commenced my experiments by placing some of them in a white-hot petroleum fire. I found that I could evaporate as much as 26 1/2 lbs. of water per square foot of heating surface per hour, and that with a forced circulation, although the quantity of water passing was very small but positive, there was no danger of overheating. I conducted many experiments with a pressure of over 400 lbs. per square inch, but none of the tubes failed. I then mounted a single tube in a white-hot furnace, also with a water circulation, and found that it only burst under steam at a pressure of 1,650 lbs. per square inch. A large boiler, having about 800 square feet of heating surface, including the feed-water heater, was then constructed. This boiler is about 4 1/2 feet wide at the bottom, 8 feet long and 6 feet high. It weighs, with the casing, the dome, and the smoke stack and connections, a little less than 1,000 lbs. The water first passes through a system of small tubes--1/4 inch in diameter and 1/60 inch thick--which were placed at the top of the boiler and immediately over the large tubes.... This feed-water heater is found to be very effective. It utilises the heat of the products of combustion after they have passed through the boiler proper and greatly reduces their temperature, while the feed-water enters the boiler at a temperature of about 250 F. A forced circulation is maintained in the boiler, the feed-water entering through a spring valve, the spring valve being adjusted in such a manner that the pressure on the water is always 30 lbs. per square inch in excess of the boiler pressure. This fall of 30 lbs. in pressure acts upon the surrounding hot water which has already passed through the tubes, and drives it down through a vertical outside tube, thus ensuring a positive and rapid circulation through all the tubes. This apparatus is found to act extremely well.'

Thus Maxim, who with this engine as power for his large aeroplane achieved free flight once, as a matter of experiment, though for what distance or time the machine was actually off the ground is matter for debate, since it only got free by tearing up the rails which were to have held it down in the experiment. Here, however, was a steam engine which was practicable for use in the air, obviously, and only the rapid success of the internal combustion engine prevented the steam-producing type from being developed toward perfection.

The first designers of internal combustion engines, knowing nothing of the petrol of these days, constructed their examples with a view to using gas as fuel. As far back as 1872 Herr Paul Haenlein obtained a speed of about 10 miles an hour with a balloon propelled by an internal combustion engine, of which the fuel was gas obtained from the balloon itself. The engine in this case was of the Lenoir type, developing some 6 horse-power, and, obviously, Haenlein's flights were purely experimental and of short duration, since he used the gas that sustained him and decreased the lifting power of his balloon with every stroke of the piston of his engine. No further progress appears to have been made with the gas-consuming type of internal combustion engine for work with aircraft; this type has the disadvantage of requiring either a gas-producer or a large storage capacity for the gas, either of which makes the total weight of the power plant much greater than that of a petrol engine. The latter type also requires less attention when working, and the fuel is more convenient both for carrying and in the matter of carburation.

The first airship propelled by the present-day type of internal combustion engine was constructed by Baumgarten and Wolfert in 1879 at Leipzig, the engine being made by Daimler with a view to working on benzine--petrol as a fuel had not then come to its own. The construction of this engine is interesting since it was one of the first of Daimler's make, and it was the development brought about by the experimental series of which this engine was one that led to the success of the motor-car in very few years, incidentally leading to that fining down of the internal combustion engine which has facilitated the development of the aeroplane with such remarkable rapidity. Owing to the faulty construction of the airship no useful information was obtained from Daimler's pioneer installation, as the vessel got out of control immediately after it was first launched for flight, and was wrecked. Subsequent attempts at mechanically-propelled flight by Wolfert ended, in 1897, in the balloon being set on fire by an explosion of benzine vapour, resulting in the death of both the aeronauts.

Daimler, from 1882 onward, devoted his attention to the perfecting of the small, high-speed petrol engine for motor-car work, and owing to his efforts, together with those of other pioneer engine-builders, the motorcar was made a success. In a few years the weight of this type of engine was reduced from near on a hundred pounds per horse-power to less than a tenth of that weight, but considerable further improvement had to be made before an engine suitable for use with aircraft was evolved.

The increase in power of the engines fitted to airships has made steady progress from the outset; Haenlein's engine developed about 6 horse-power; the Santos-Dumont airship of 1898 was propelled by a motor of 4 horse-power; in 1902 the

Lebaudy airship was fitted with an engine of 40 horse-power, while, in 1910, the Lebaudy brothers fitted an engine of nearly 300 horsepower to the airship they were then constructing--1,400 horse-power was common in the airships of the War period, and the later British rigids developed yet more.

Before passing on to consideration of the petrol-driven type of engine, it is necessary to accord brief mention to the dirigible constructed in 1884 by Gaston and Albert Tissandier, who at Grenelle, France, achieved a directed flight in a wind of 8 miles an hour, obtaining their power for the propeller from 1 1/3 horse-power Siemens electric motor, which weighed 121 lbs. and took its current from a bichromate battery weighing 496 lbs. A two-bladed propeller, 9 feet in diameter, was used, and the horse-power output was estimated to have run up to 1 1/2 as the dirigible successfully described a semicircle in a wind of 8 miles an hour, subsequently making headway transversely to a wind of 7 miles an hour. The dirigible with which this motor was used was of the conventional pointed-end type, with a length of 92 feet, diameter of 30 feet, and capacity of 37,440 cubic feet of gas. Commandant Renard, of the French army balloon corps, followed up Tissandier's attempt in the next year--1885--making a trip from Chalais-Meudon to Paris and returning to the point of departure quite successfully. In this case the motive power was derived from an electric plant of the type used by the Tissandiers, weighing altogether 1,174 lbs., and developing 9 horsepower. A speed of 14 miles an hour was attained with this dirigible, which had a length of 165 feet, diameter of 27 feet, and capacity of 65,836 cubic feet of gas.

Reverting to the petrol-fed type again, it is to be noted that Santos-Dumont was practically the first to develop the use of the ordinary automobile engine for air work--his work is of such importance that it has been considered best to treat of it as one whole, and details of the power plants are included in the account of his experiments. Coming to the Lebaudy brothers and their work, their engine of 1902 was a 40 horse-power Daimler, four-cylindered; it was virtually a large edition of the Daimler car engine, the arrangement of the various details being on the lines usually adopted for the standard Daimler type of that period. The cylinders were fully water-jacketed, and no special attempt toward securing lightness for air work appears to have been made.

The fining down of detail that brought weight to such limits as would fit the engine for work with heavier-than-air craft appears to have waited for the brothers Wright. Toward the end of 1903 they fitted to their first practicable flying machine the engine which made the historic first aeroplane flight; this engine developed 30 horse-power, and weighed only about 7 lbs. per horse-power developed, its design and workmanship being far ahead of any previous design in this respect, with the exception of the remarkable engine, designed by Manly, installed in Langley's ill-fated aeroplane--or 'aerodrome,' as he preferred to call it--tried in 1903.

The light weight of the Wright brothers' engine did not necessitate a high number of revolutions per minute to get the requisite power; the speed was only 1,300 revolutions per minute, which, with a piston stroke of 3.94 inches, was quite moderate. Four cylinders were used, the cylinder diameter being 4.42 inches; the engine was of the vertical type, arranged to drive two propellers at a rate of about 350 revolutions per minute, gearing being accomplished by means of chain drive from crank-shaft end to propeller spindle.

The methods adopted by the Wrights for obtaining a light-weight engine were of considerable interest, in view of the fact that the honour of first achieving flight by means of the driven plane belongs to them--unless Ader actually flew as he claimed. The cylinders of this first Wright engine were separate castings of steel, and only the barrels were jacketed, this being done by fixing loose, thin aluminium covers round the outside of each cylinder. The combustion head and valve pockets were cast together with the cylinder barrel, and were not water cooled. The inlet valves were of the automatic type, arranged on the tops of the cylinders, while the exhaust valves were also overhead, operated by rockers and push-rods. The pistons and piston rings were of the ordinary type, made of cast-iron, and the connecting rods were circular in form, with a hole drilled down the middle of each to reduce the weight.

Necessity for increasing power and ever lighter weight in relation to the power produced has led to the evolution of a number of different designs of internal combustion engines. It was quickly realised that increasing the number of cylinders on an engine was a better way of getting more power than that of increasing the cylinder diameter, as the greater number of cylinders gives better torque-even turning effect--as well as keeping down the weight--this latter because the bigger cylinders must be more stoutly constructed than the small sizes; this fact has led to the construction of engines having as many as eighteen cylinders, arranged in three parallel rows in order to keep the length of crankshaft within reasonable limits. The aero engine of to-day may, roughly, be divided into four classes: these are the V type, in which two rows of cylinders are set parallel at a certain angle to each other; the radial type, which consists of cylinders arranged radially and remaining stationary while the crankshaft revolves; the rotary, where the cylinders are disposed round a common centre and revolve round a stationary shaft, and the vertical type, of four or six cylinders--seldom more than this--arranged in one row. A modification of the V type is the eighteen-cylindered engine--the Sunbeam is one of the best examples--in which three rows of cylinders are set parallel to each other, working on a common crankshaft. The development these four types started with that of the vertical--the simplest of all; the V, radial, and rotary types came after the vertical, in the order given.

The evolution of the motor-car led to the adoption of the vertical type of internal combustion engine in preference to any other, and it followed naturally that vertical engines should be first used for aeroplane propulsion, as by taking an engine that had been developed to some extent, and adapting it to its new work, the problem of mechanical flight was rendered easier than if a totally new type had had to be evolved. It was quickly realised--by the Wrights, in fact-that the minimum of weight per horse-power was the prime requirement for the successful development of heavier-than-air machines, and at the same time it was equally apparent that the utmost reliability had to be obtained from the engine, while a third requisite was economy, in order to reduce the weight of petrol necessary for flight.

Daimler, working steadily toward the improvement of the internal combustion engine, had made considerable progress by the end of last century. His two-cylinder engine of 1897 was approaching to the present-day type, except as regards the method of ignition; the cylinders had 3.55 inch diameter, with a 4.75 inch piston stroke, and the engine was rated at 4.5 brake horse-power, though it probably developed more than this in actual running at its rated speed of 800 revolutions per minute. Power was limited by the inlet and exhaust passages, which, compared with present-day practice, were very small. The heavy castings of which the engine was made up are accounted for by the necessity for considering foundry practice of the time, for in 1897 castings were far below the present-day standard. The crank-case of this two-cylinder vertical Daimler engine was the only part made of aluminium, and even with this no attempt was made to attain lightness, for a circular flange was cast at the bottom to form a stand for the engine during machining and erection. The general design can be followed from the sectional views, and these will show, too, that ignition was by means of a hot tube on the cylinder head, which had to be heated with a blow-lamp before starting the engine. With all its well known and hated troubles, at that time tube ignition had an advantage over the magneto, and the coil and accumulator system, in reliability; sparking plugs, too, were not so reliable then as they are now. Daimler fitted a very simple type of carburettor to this engine, consisting only of a float with a single jet placed in the air passage. It may be said that this twin-cylindered vertical was the first of the series from which has been evolved the Mercedes-Daimler car and airship engines, built in sizes up to and even beyond 240 horse-power.

In 1901 the development of the petrol engine was still so slight that it did not admit of the construction, by any European maker, of an engine weighing less than 12 lbs. per horse-power. Manly, working at the instance of Professor Langley, produced a five-cylindered radial type engine, in which both the design and workmanship showed a remarkable advance in construction. At 950 revolutions per minute it developed 52.4 horse-power, weighing only 2.4 pounds per horse-power; it was a very remarkable achievement in engine design, considering the power developed in relation to the total weight, and it was, too, an interruption in the development of the vertical type which showed that there were other equally great possibilities in design.

In England, the first vertical aero-engine of note was that designed by Green, the cylinder dimensions being 4.15 inch diameter by 4.75 stroke--a fairly complete idea of this engine can be obtained from the accompanying diagrams. At a speed of 1,160 revolutions per minute it developed 35 brake horse-power, and by accelerating up to 1,220 revolutions per minute a maximum of 40 brake horse-power could be obtained--the first-mentioned was the rated working speed of the engine for continuous runs. A flywheel, weighing 23.5 lbs., was fitted to the engine, and this, together with the ignition system, brought the weight up to 188 lbs., giving 5.4 lbs. per horse-power. In comparison with the engine fitted to the Wrights' aeroplane a greater power was obtained from approximately the same cylinder volume, and an appreciable saving in weight had also been effected. The illustration shows the arrangement of the vertical valves at the top of the cylinder and the overhead cam shaft, while the position of the carburettor and inlet pipes can be also seen. The water jackets were formed by thin copper casings, each cylinder being separate and having its independent jacket rigidly fastened to the cylinder at the top only, thus allowing for free expansion of the casing; the joint at the bottom end was formed by sliding the jacket over a rubber ring. Each cylinder was bolted to the crank-case and set out of line with the crankshaft, so that the crank has passed over the upper dead centre by the time that the piston is at the top of its stroke when receiving the full force of fuel explosion. The advantage of this desaxe setting is that the pressure in the cylinder acts on the crank-pin with a more effective leverage during that part of the stroke when that pressure is highest, and in addition the side pressure of the piston on the cylinder wall, due to the thrust of the connecting rod, is reduced. Possibly the charging of the cylinder is also more complete by this arrangement, owing to the slower movement of the piston at the bottom of its stroke allowing time for an increased charge of mixture to enter the cylinder.

A 60 horse-power engine was also made, having four vertical cylinders, each with a diameter of 5.5 inches and stroke of 5.75 inches, developing its rated power at 1,100 revolutions per minute. By accelerating up to 1,200 revolutions per minute 70 brake horsepower could be obtained, and a maximum of 80 brake horse-power was actually attained with the type. The flywheel, fitted as with the original 35 horse-power engine, weighed 37 lbs.; with this and with the ignition system the total weight of the engine was only 250 lbs., or 4.2 lbs. per horse-power at the normal rating. In this design, however, low weight in relation to power was not the ruling factor, for Green gave more attention to reliability and economy of fuel consumption, which latter was approximately 0.6 pint of petrol per brake horse-power per hour. Both the oil for lubricating the bearings and the water for cooling the cylinders were circulated by pumps, and all parts of the valve gear, etc., were completely enclosed for protection from dust.

A later development of the Green engine was a six-cylindered vertical, cylinder dimensions being 5.5 inch diameter by 6 inch stroke, developing 120 brake horsepower when running at 1,250 revolutions per minute. The total weight of the engine with ignition system 398 was 440 lbs., or 3.66 lbs. per horse-power. One of these engines was used on the machine which, in 1909, won the prize of L1,000 for the first circular mile flight, and it may be noted, too, that S. F. Cody, making the circuit of England in 1911, used a four-cylinder Green engine. Again, it was a Green engine that in 1914 won the L5,000 prize offered for the best aero engine in the Naval and Military aeroplane engine competition.

Manufacture of the Green engines, in the period of the War, had standardised to the production of three types. Two of these were six-cylinder models, giving respectively 100 and 150 brake horse-power, and the third was a twelve-cylindered model rated at 275 brake horse-power.

In 1910 J. S. Critchley compiled a list showing the types of engine then being manufactured; twenty-two out of a total of seventy-six were of the four-cylindered vertical type, and in addition to these there were two six-cylindered verticals. The sizes of the four-cylinder types ranged from 26 up to 118 brake horse-power; fourteen of them developed less than 50 horse-power, and only two developed over 100 horse-power.

It became apparent, even in the early stages of heavier-than-air flying, that four-cylinder engines did not produce the even torque that was required for the rotation of the power shaft, even though a flywheel was fitted to the engine. With this type of engine the breakage of air-screws was of frequent occurrence, and an engine having a more regular rotation was sought, both for this and to avoid the excessive vibration often experienced with the four-cylinder type. Another, point that forced itself on engine builders was that the increased power which was becoming necessary for the propulsion of aircraft made an increase in the number of cylinders essential, in order to obtain a light engine. An instance of the weight reduction obtainable in using six cylinders instead of four is shown in Critchley's list, for one of the four-cylinder engines developed 118.5 brake horse-power and weighed 1,100 lbs., whereas a six-cylinder engine by the same manufacturer developed 117.5 brake horse-power with a weight of 880 lbs., the respective cylinder dimensions being 7.48 diameter by 9.06 stroke for the four-cylinder engine, and 6.1 diameter by 7.28 stroke for the six-cylinder type.

A list of aeroplane engines, prepared in 1912 by Graham Clark, showed that, out of the total number of 112 engines then being manufactured, forty-two were of the vertical type, and of this number twenty-four had four-cylinders while sixteen were six-cylindered. The German aeroplane engine trials were held a year later, and sixty-six engines entered the competition, fourteen of these being made with air-cooled cylinders. All of the ten engines that were chosen for the final trials were of the water-cooled type, and the first place was won by a Benz four-cylinder vertical engine which developed 102 brake horse-power at 1,288 revolutions per minute. The cylinder dimensions of this engine were 5.1 inch diameter by 7.1 inch stroke, and the weight of the engine worked out at 3.4 lbs. per brake horse-power. During the trials the full-load petrol consumption was 0.53 pint per horse-power per hour, and the amount of lubricating oil used was 0.0385 pint per brake horse-power per hour. In general construction this Benz engine was somewhat similar to the Green engine already described; the overhead valves, fitted in the tops of the cylinders, were similarly arranged, as was the cam-shaft; two springs were fitted to each of the valves to guard against the possibility of the engine being put out of action by breakage of one of the springs, and ignition was obtained by two high-tension magnetos giving simultaneous sparks in each cylinder by means of two sparking plugs--this dual ignition reduced the possibility of ignition troubles. The cylinder jackets were made of welded sheet steel so fitted around the cylinder that the head was also water-cooled, and the jackets were corrugated in the middle to admit of independent expansion. Even the lubrication system was duplicated, two sets of pumps being used, one to circulate the main supply of lubricating oil, and the other to give a continuous supply of fresh oil to the bearings, so that if the supply from one pump failed the other could still maintain effective lubrication.

Development of the early Daimler type brought about the four-cylinder vertical Mercedes-Daimler engine of 85 horse-power, with cylinders of 5.5 diameter with 5.9 inch stroke, the cylinders being cast in two pairs. The overhead arrangement of valves was adopted, and in later designs push-rods were eliminated, the overhead cam-shaft being adopted in their place. By 1914 the four-cylinder Mercedes-Daimler had been partially displaced from favour by a six-cylindered model, made in two sizes; the first of these gave a nominal brake horse-power of 80, having cylinders of 4.1 inches diameter by 5.5 inches stroke; the second type developed 100 horse-power with cylinders 4.7 inches in diameter and 5.5 inches stroke, both types being run at 1,200 revolutions per minute. The cylinders of both these types were cast in pairs, and, instead of the water jackets forming part of the casting, as in the design of the original four-cylinder Mercedes-Daimler engine, they were made of steel welded to flanges on the cylinders. Steel pistons, fitted with cast-iron rings, were used, and the overhead arrangement of valves and cam-shaft was adopted. About 0.55 pint per brake horse-power per hour was the usual fuel consumption necessary to full load running, and the engine was also economical as regards the consumption of lubricating oil, the lubricating system being 'forced' for all parts, including the cam-shaft. The shape of these engines was very well suited for work with aircraft, being narrow enough to admit of a streamline form being obtained, while all the accessories could be so mounted as to produce little or no wind resistance, and very little obstruction to the pilot's view.

The eight-cylinder Mercedes-Daimler engine, used for airship propulsion during the War, developed 240 brake horse-power at 1,100 revolutions per minute; the cylinder dimensions were 6.88 diameter by 6.5 stroke--one of the instances in which the short stroke in relation to bore was very noticeable.

Other instances of successful vertical design-the types already detailed are fully sufficient to give particulars of the type generally--are the Panhard, Chenu, Maybach, N.A.G., Argus, Mulag, and the well-known Austro-Daimler, which by 1917 was being copied in every combatant country. There are also the later Wright engines, and in America the Wisconsin six-cylinder vertical, weighing well under 4 lbs. per horse-power, is evidence of the progress made with this first type of aero engine to develop.

II. THE VEE TYPE

An offshoot from the vertical type, doubling the power of this with only a very slight--if any--increase in the length of crankshaft, the Vee or diagonal type of aero engine leaped to success through the insistent demand for greater power. Although the design came after that of the vertical engine, by 1910, according to Critchley's list of aero engines, there were more Vee type engines being made than any other type, twenty-five sizes being given in the list, with an average rating of 57.4 brake horse-power.

The arrangement of the cylinders in Vee form over the crankshaft, enabling the pistons of each pair of opposite cylinders to act upon the same crank pin, permits of a very short, compact engine being built, and also permits of reduction of the weight per horsepower, comparing this with that of the vertical type of engine, with one row of cylinders. Further, at the introduction of this type of engine it was seen that crankshaft vibration, an evil of the early vertical engines, was practically eliminated, as was the want of longitudinal stiffness that characterised the higher-powered vertical engines.

Of the Vee type engines shown in Critchley's list in 1910 nineteen different sizes were constructed with eight cylinders, and with horse-powers ranging from thirty to just over the hundred; the lightest of these weighed 2.9 lbs. per horsepower--a considerable advance in design on the average vertical engine, in this respect of weight per horse-power. There were also two sixteen-cylinder engines of Vee design, the larger of which developed 134 horse-power with a weight of only 2 lbs. per brake horse-power. Subsequent developments have indicated that this type, with the further development from it of the double-Vee, or engine with three rows of cylinders, is likely to become the standard design of aero engine where high powers are required. The construction permits of placing every part so that it is easy of access, and the form of the engine implies very little head resistance, while it can be placed on the machine--supposing that machine to be of the single-engine type--in such a way that the view of the pilot is very little obstructed while in flight.

An even torque, or great uniformity of rotation, is transmitted to the air-screw by these engines, while the design also permits of such good balance of the engine itself that vibration is practically eliminated. The angle between the two rows of cylinders is varied according to the number of cylinders, in order to give working impulses at equal angles of rotation and thus provide even torque; this angle is determined by dividing the number of degrees in a circle by the number of cylinders in either row of the engine. In an eight-cylindered Vee type engine, the angle between the cylinders is 90 degrees; if it is a twelve-cylindered engine, the angle drops to 60 degrees.

One of the earliest of the British-built Vee type engines was an eight-cylinder 50 horse-power by the Wolseley Company, constructed in 1908 with a cylinder bore of 3.75 inches and stroke of 5 inches, running at a normal speed of 1,350 revolutions per minute. With this engine, a gearing was introduced to enable the propeller to run at a lower speed than that of the engine, the slight loss of efficiency caused by the friction of the gearing being compensated by the slower speed of the air-screw, which had higher efficiency than would have been the case if it had been run at the engine speed. The ratio of the gearing--that is, the speed of the air-screw relatively to that of the engine, could be chosen so as to suit exactly the requirements of the air-screw, and the gearing itself, on this engine, was accomplished on the half-speed shaft actuating the valves.

Very soon after this first design had been tried out, a second Vee type engine was produced which, at 1,200 revolutions per minute, developed 60 horse-power; the size of this engine was practically identical with that of its forerunner, the only exception being an increase of half an inch in the cylinder stroke--a very long stroke of piston in relation to the bore of the cylinder. In the first of these two engines, which was designed for airship propulsion, the weight had been about 8 lbs. per brake horse-power, no special attempt appearing to have been made to fine down for extreme lightness; in this 60 horse-power design, the weight was reduced to 6.1 lbs. per horse-power, counting the latter as normally rated; the engine actually gave a maximum of 75 brake horse-power, reducing the ratio of weight to power very considerably below the figure given.

The accompanying diagram illustrates a later Wolseley model, end elevation, the eight-cylindered 120 horse-power Vee type aero engine of the early war period. With this engine, each crank pin has two connecting rods bearing on it, these being placed side by side and connected to the pistons of opposite cylinders and the two cylinders of the pair are staggered by an amount equal to the width of the connecting rod bearing, to afford accommodation for the rods. The crankshaft was a nickel chrome steel forging, machined hollow, with four crank pins set at 180 degrees to each other, and carried in three bearings lined with anti-friction metal. The connecting rods were made of tubular nickel chrome steel, and the pistons of drawn steel, each being fitted with four piston rings. Of these the two rings nearest to the piston head were of the ordinary cast-iron type, while the others were of phosphor bronze, so arranged as to take the side thrust of the piston. The cylinders were of steel, arranged in two groups or rows of four, the angular distance between them being 90 degrees. In the space above the crankshaft, between the cylinder rows, was placed the valve-operating mechanism, together with the carburettor and ignition system, thus rendering this a very compact and accessible engine. The combustion heads of the cylinders were made of cast-iron, screwed into the steel cylinder barrels; the water-jacket was of spun aluminium, with one end fitting over the combustion head and the other free to slide on the cylinder; the water-joint at the lower end was made tight by a Dermatine ring carried between small flanges formed on the cylinder barrel. Overhead valves were adopted, and in order to make these as large as possible the combustion chamber was made slightly larger in diameter than the cylinder, and the valves set at an angle. Dual ignition was fitted in each cylinder, coil and accumulator being used for starting and as a reserve in case of failure of the high-tension magneto system fitted for normal running. There was a double set of lubricating pumps, ensuring continuity of the oil supply to all the bearings of the engine.

The feature most noteworthy in connection with the running of this type of engine was its flexibility; the normal output of power was obtained with 1,150 revolutions per minute of the crankshaft, but, by accelerating up to 1,400 revolutions, a maximum of 147 brake horse-power could be obtained. The weight was about 5 lbs. per horse-power, the cylinder dimensions being 5 inches bore by 7 inches stroke. Economy in running was obtained, the fuel consumption being 0.58 pint per brake horse-power per hour at full load, with an expenditure of about 0.075 pint of lubricating oil per brake horse-power per hour.

Another Wolseley Vee type that was standardised was a 90 horse-power eight-cylinder engine running at 1,800 revolutions per minute, with a reducing gear introduced by fitting the air screw on the half-speed shaft. First made semi-cooled--the exhaust valve was left air-cooled, and then entirely water-jacketed--this engine demonstrated the advantage of full water cooling, for under the latter condition the same power was developed with cylinders a quarter of an inch less in diameter than in the semi-cooled pattern; at the same time the weight was brought down to 4 1/2 lbs. per horsepower.

A different but equally efficient type of Vee design was the Dorman engine, of which an end elevation is shown; this developed 80 brake horse-power at a speed of 1,300 revolutions per minute, with a cylinder bore of 5 inches; each cylinder was made in cast-iron in one piece with the combustion chamber, the barrel only being water-jacketed. Auxiliary exhaust ports were adopted, the holes through the cylinder wall being uncovered by the piston at the bottom of its stroke--the piston, 4.75 inches in length, was longer than its stroke, so that these ports were covered when it was at the top of the cylinder. The exhaust discharged through the ports into a belt surrounding the cylinder, the belts on the cylinders being connected so that the exhaust gases were taken through a single pipe. The air was drawn through the crank case, before reaching the carburettor, this having the effect of cooling the oil in the crank case as well as warming the air and thus assisting in vaporising the petrol for each charge of the cylinders. The inlet and exhaust valves were of the overhead type, as may be gathered from the diagram, and in spite of cast-iron cylinders being employed a light design was obtained, the total weight with radiator, piping, and water being only 5.5 lbs. per horse-power.

Here was the antithesis of the Wolseley type in the matter of bore in relation to stroke; from about 1907 up to the beginning of the war, and even later, there was controversy as to which type--that in which the bore exceeded the stroke, or vice versa--gave greater efficiency. The short-stroke enthusiasts pointed to the high piston speed of the long-stroke type, while those who favoured the latter design contended that full power could not be obtained from each explosion in the short-stroke type of cylinder. It is now generally conceded that the long-stroke engine yields higher efficiency, and in addition to this, so far as car engines are concerned, the method of rating horse-power in relation to bore without taking stroke into account has given the long-stroke engine an advantage, actual horse-power with a long stroke engine being in excess of the nominal rating. This may have had some influence on aero engine design, but, however this may have been, the long-stroke engine has gradually come to favour, and its rival has taken second place.

For some time pride of place among British Vee type engines was held by the Sunbeam Company, which, owing to the genius of Louis Coatalen, together with the very high standard of construction maintained by the firm, achieved records and fame in the middle and later periods of the war. Their 225 horse-power twelve-cylinder engine ran at a normal speed of 2,000 revolutions per minute; the air screw was driven through gearing at half this speed, its shaft being separate from the timing gear and carried in ball-bearings on the nose-piece of the engine. The cylinders were of cast-iron, entirely water-cooled; a thin casing formed the water-jacket, and a very light design was obtained, the weight being only 3.2 lbs. per horse-power.

The first engine of Sunbeam design had eight cylinders and developed 150 horsepower at 2,000 revolutions per minute; the final type of Vee design produced during the war was twelve-cylindered, and yielded 310 horse-power with cylinders 4.3 inches bore by 6.4 inches stroke. Evidence in favour of the long-stroke engine is afforded in this type as regards economy of working; under full load, working at 2,000 revolutions per minute, the consumption was 0.55 pints of fuel per brake horse-power per hour, which seems to indicate that the long stroke permitted of full use being made of the power resulting from each explosion, in spite of the high rate of speed of the piston.

Developing from the Vee type, the eighteen-cylinder 475 brake horse-power engine, designed during the war, represented for a time the limit of power obtainable from a single plant. It was water-cooled throughout, and the ignition to each cylinder was duplicated; this engine proved fully efficient, and economical in fuel consumption. It was largely used for seaplane work, where reliability was fully as necessary as high power.

The abnormal needs of the war period brought many British firms into the ranks of Vee-type engine-builders, and, apart from those mentioned, the most notable types produced are the Rolls-Royce and the Napier. The first mentioned of these firms, previous to 1914 had concentrated entirely on car engines, and their very high standard of production in this department of internal combustion engine work led, once they took up the making of aero engines, to extreme efficiency both of design and workmanship. The first experimental aero engine, of what became known as the 'Eagle' type, was of Vee design--it was completed in March of 1915-- and was so successful that it was standardised for quantity production. How far the original was from the perfection subsequently ascertained is shown by the steady increase in developed horse-power of the type; originally designed to develop 200 horse-power, it was developed and improved before its first practical trial in October of 1915, when it developed 255 horsepower on a brake test. Research and experiment produced still further improvements, for, without any enlargement of the dimensions, or radical alteration in design, the power of the engine was brought up to 266 horse-power by March of 1916, the rate of revolutions of 1,800 per minute being maintained throughout. July, 1916 gave 284 horse-power; by the cud of the year this had been increased to 322 horse-power; by September of 1917 the increase was to 350 horse-power, and by February of 1918 then 'Eagle' type of engine was rated at 360 horse-power, at which standard it stayed. But there is no more remarkable development in engine design than this, a 75 per cent increase of power in the same engine in a period of less than three years.

To meet the demand for a smaller type of engine for use on training machines, the Rolls-Royce firm produced the 'Hawk' Vee-type engine of 100 horsepower, and, intermediately between this and the 'Eagle,' the 'Falcon' engine came to being with an original rated horse-power of 205 at 1,800 revolutions per minute, in April of 1916. Here was another case of growth of power in the same engine through research, almost similar to that of the 'Eagle' type, for by July of 1918 the 'Falcon' was developing 285 horse-power with no radical alteration of design. Finally, in response to the constant demand for increase of power in a single plant, the Rolls-Royce company designed and produced the 'Condor' type of engine, which yielded 600 horse-power on its first test in August of 1918. The cessation of hostilities and consequent falling off in the demand for extremely high-powered plants prevented the 'Condor' being developed to its limit, as had been the 'Falcon' and 'Eagle' types.

The 'Eagle 'engine was fitted to the two Handley-Page aeroplanes--which made flights from England to India--it was virtually standard on the Handley-Page bombers of the later War period, though to a certain extent the American 'Liberty' engine was also used. Its chief record, however, is that of being the type fitted to the Vickers-Vimy aeroplane which made the first Atlantic flight, covering the distance of 1,880 miles at a speed averaging 117 miles an hour.

The Napier Company specialised on one type of engine from the outset, a power plant which became known as the 'Lion' engine, giving 450 horse-power with twelve cylinders arranged in three rows of four each. Considering the engine as 'dry,' or without fuel and accessories, an abnormally light weight per horse-power--only 1.89 lbs.--was attained when running at the normal rate of revolution. The cylinders and water-jackets are of steel, and there is fitted a detachable aluminium cylinder head containing inlet and exhaust valves and valve actuating mechanism; pistons are of aluminium alloy, and there are two inlet and two exhaust valves to each cylinder, the whole of the valve mechanism being enclosed in an oil-tight aluminium case. Connecting rods and crankshaft are of steel, the latter being machined from a solid steel forging and carried in five roller bearings and one plain bearing at the forward end. The front end of the crank-case encloses reduction gear for the propeller shaft, together with the shaft and bearings. There are two suction and one pressure type oil pumps driven through gears at half-engine speed, and two 12 spark magnetos, giving 2 sparks in each cylinder.

The cylinders are set with the central row vertical, and the two side rows at angles of 60 degrees each; cylinder bore is 5 1/2 inches, and stroke 5 1/8 inches; the normal rate of revolution is 1,350 per minute, and the reducing gear gives one revolution of the propeller shaft to 1.52 revolutions of crankshaft. Fuel consumption is 0.48lbs. of fuel per brake horse-power hour at full load, and oil consumption is 0.020 lbs. per brake horsepower hour. The dry weight of the engine, complete with propeller boss, carburettors, and induction pipes, is 850 lbs., and the gross weight in running order, with fuel and oil for six hours working, is 2,671 lbs., exclusive of cooling water.

To this engine belongs an altitude record of 30,500 feet, made at Martlesham, near Ipswich, on January 2nd, 1919, by Captain Lang, R.A.F., the climb being accomplished in 66 minutes 15 seconds. Previous to this, the altitude record was held by an Italian pilot, who made 25,800 feet in an hour and 57 minutes in 1916. Lang's climb was stopped through the pressure of air, at the altitude he reached, being insufficient for driving the small propellers on the machine which worked the petrol and oil pumps, or he might have made the height said to have been attained by Major Schroeder on February 27th, 1920, at Dayton, Ohio. Schroeder is said to have reached an altitude of 36,020 feet on a Napier biplane, and, owing to failure of the oxygen supply, to have lost consciousness, fallen five miles, righted his machine when 2,000 feet in the air, and alighted successfully. Major Schroeder is an American.

Turning back a little, and considering other than British design of Vee and double-Vee or 'Broad arrow' type of engine, the Renault firm from the earliest days devoted considerable attention to the development of this type, their air-cooled engines having been notable examples from the earliest days of heavier-than-air machines. In 1910 they were making three sizes of eight-cylindered Vee-type engines, and by 1915 they had increased to the manufacture of five sizes, ranging from 25 to 100 brake horse-power, the largest of the five sizes having twelve cylinders but still retaining the air-cooled principle. The De Dion firm, also, made Vee-type engines in 1914, being represented by an 80 horse-power eight-cylindered engine, air-cooled, and a 150 horse-power, also of eight cylinders, water-cooled, running at a normal rate of 1,600 revolutions per minute. Another notable example of French construction was the Panhard and Levassor 100 horse-power eight-cylinder Vee engine, developing its rated power at 1,500 revolutions per minute, and having the--for that time--low weight of 4.4 lbs. per horse-power.

American Vee design has followed the British fairly cclosely; the Curtiss Company produced originally a 75 horse-power eight-cylinder Vee type running at 1,200 revolutions per minute, supplementing this with a 170 horse-power engine running at 1,600 revolutions per minute, and later with a twelve-cylinder model Vee type, developing 300 horse-power at 1,500 revolutions per minute, with cylinder bore of 5 inches and stroke of 7 inches. An exceptional type of American design was the Kemp Vee engine of 80 horse-power in which the cylinders were cooled by a current of air obtained from a fan at the forward end of the engine. With cylinders of 4.25 inches bore and 4.75 inches stroke, the rater power was developed at 1,150 revolutions per minute, and with the engine complete the weight was only 4.75 lbs. per horse-power.

III. THE RADIAL TYPE

The very first successful design of internal combustion aero engine made was that of Charles Manly, who built a five-cylinder radial engine in 1901 for use with Langley's 'aerodrome,' as the latter inventor decided to call what has since become known as the aeroplane. Manly made a number of experiments, and finally decided on radial design, in which the cylinders are so rayed round a central crank-pin that the pistons act successively upon it; by this arrangement a very short and compact engine is obtained, with a minimum of weight, and a regular crankshaft rotation and perfect balance of inertia forces.

When Manly designed his radial engine, high speed internal combustion engines were in their infancy, and the difficulties in construction can be partly realised when the lack of manufacturing methods for this high-class engine work, and the lack of experimental data on the various materials, are taken into account. During its tests, Manly's engine developed 52.4 brake horsepower at a speed of 950 revolutions per minute, with the remarkably low weight of only 2.4 lbs. per horsepower; this latter was increased to 3.6 lbs. when the engine was completed by the addition of ignition system, radiator, petrol tank, and all accessories, together with the cooling water for the cylinders.

In Manly's engine, the cylinders were of steel, machined outside and inside to 1/16 of an inch thickness; on the side of cylinder, at the top end, the valve chamber was brazed, being machined from a solid forging, The casing which formed the water-jacket was of sheet steel, 1/50 of an inch in thickness, and this also was brazed on the cylinder and to the valve chamber. Automatic inlet valves were fitted, and the exhaust valves were operated by a cam which had two points, 180 degrees apart; the cam was rotated in the opposite direction to the engine at one-quarter engine speed. Ignition was obtained by using a one-spark coil and vibrator for all cylinders, with a distributor to select the right cylinder for each spark--this was before the days of the high-tension magneto and the almost perfect ignition systems that makers now employ. The scheme of ignition for this engine was originated by Manly himself, and he also designed the sparking plugs fitted in the tops of the cylinders. Through fear of trouble resulting if the steel pistons worked on the steel cylinders, cast iron liners were introduced in the latter, 1/16 of an inch thick.

The connecting rods of this engine were of virtually the same type as is employed on nearly all modern radial engines. The rod for one cylinder had a bearing along the whole of the crank pin, and its end enclosed the pin; the other four rods had bearings upon the end of the first rod, and did not touch the crank pin. The accompanying diagram shows this construction, together with the means employed for securing the ends of the four rods--the collars were placed in position after the rods had been put on. The bearings of these rods did not receive any of the rubbing effect due to the rotation of the crank pin, the rubbing on them being only that of the small angular displacement of the rods during each revolution; thus there was no difficulty experienced with the lubrication.

Another early example of the radial type of engine was the French Anzani, of which type one was fitted to the machine with which Bleriot first crossed the English Channel--this was of 25 horse-power. The earliest Anzani engines were of the three-cylinder fan type, one cylinder being vertical, and the other two placed at an angle of 72 degrees on each side, as the possibility of over-lubrication of the bottom cylinders was feared if a regular radial construction were adopted. In order to overcome the unequal balance of this type, balance weights were fitted inside the crank case.

The final development of this three-cylinder radial was the 'Y' type of engine, in which the cylinders were regularly disposed at 120 degrees apart, the bore was 4.1, stroke 4.7 inches, and the power developed was 30 brake horse-power at 1,300 revolutions per minute.

Critchley's list of aero engines being constructed in 1910 shows twelve of the radial type, with powers of between 14 and 100 horse-power, and with from three to ten cylinder--this last is probably the greatest number of cylinders that can be successfully arranged in circular form. Of the twelve types of 1910, only two were water-cooled, and it is to be noted that these two ran at the slowest speeds and had the lowest weight per horse-power of any.

The Anzani radial was considerably developed special attention being paid to this type by its makers and by 1914 the Anzani list comprised seven different sizes of air-cooled radials. Of these the largest had twenty cylinders, developing 200 brake horse-power--it was virtually a double radial--and the smallest was the original 30 horse-power three-cylinder design. A six-cylinder model was formed by a combination of two groups of three cylinders each, acting upon a double-throw crankshaft; the two crank pins were set at 180 degrees to each other, and the cylinder groups were staggered by an amount equal to the distance between the centres of the crank pins. Ten-cylinder radial engines are made with two groups of five cylinders acting upon two crank pins set at 180 degrees to each other, the largest Anzani 'ten' developed 125 horsepower at 1,200 revolutions per minute, the ten cylinders being each 4.5 inches in bore with stroke of 5.9 inches, and the weight of the engine being 3.7 lbs. per horse-power. In the 200 horse-power Anzani radial the cylinders are arranged in four groups of five each, acting on two crank pins. The bore of the cylinders in this engine is the same as in the three-cylinder, but the stroke is increased to 5.5 inches. The rated power is developed at 1,300 revolutions per minute, and the engine complete weighs 3.4 lbs. per horse-power.

With this 200 horse-power Anzani, a petrol consumption of as low as 0.49 lbs.

of fuel per brake horse-power per hour has been obtained, but the consumption of lubricating oil is compensatingly high, being up to one-fifth of the fuel used. The cylinders are set desaxe with the crank shaft, and are of cast-iron, provided with radiating ribs for air-cooling; they are attached to the crank case by long bolts passing through bosses at the top of the cylinders, and connected to other bolts at right angles through the crank case. The tops of the cylinders are formed flat, and seats for the inlet and exhaust valves are formed on them. The pistons are cast-iron, fitted with ordinary cast-iron spring rings. An aluminium crank case is used, being made in two halves connected together by bolts, which latter also attach the engine to the frame of the machine. The crankshaft is of nickel steel, made hollow, and mounted on ball-bearings in such a manner that practically a combination of ball and plain bearings is obtained; the central web of the shaft is bent to bring the centres of the crank pins as close together as possible, leaving only room for the connecting rods, and the pins are 180 degrees apart. Nickel steel valves of the cone-seated, poppet type are fitted, the inlet valves being automatic, and those for the exhaust cam-operated by means of push-rods. With an engine having such a number of cylinders a very uniform rotation of the crankshaft is obtained, and in actual running there are always five of the cylinders giving impulses to the crankshaft at the same time.

An interesting type of pioneer radial engine was the Farcot, in which the cylinders were arranged in a horizontal plane, with a vertical crankshaft which operated the air-screw through bevel gearing. This was an eight-cylinder engine, developing 64 horse-power at 1,200 revolutions per minute. The R.E.P. type, in the early days, was a 'fan' engine, but the designer, M. Robert Pelterie, turned from this design to a seven-cylinder radial, which at 1,100 revolutions per minute gave 95 horse-power. Several makers entered into radial engine development in the years immediately preceding the War, and in 1914 there were some twenty-two different sizes and types, ranging from 30 to 600 horse-power, being made, according to report; the actual construction of the latter size at this time, however, is doubtful.

Probably the best example of radial construction up to the outbreak of War was the Salmson (Canton-Unne) water-cooled, of which in 1914 six sizes were listed as available. Of these the smallest was a seven-cylinder 90 horse-power engine, and the largest, rated at 600 horse-power, had eighteen cylinders. These engines, during the War, were made under license by the Dudbridge Ironworks in Great Britain.

The accompanying diagram shows the construction of the cylinders in the 200 horse-power size, showing the method of cooling, and the arrangement of the connecting rods. A patent planetary gear, also shown in the diagram, gives exactly the same stroke to all the pistons. The complete engine has fourteen cylinders, of forged steel machined all over, and so secured to the crank case that any one can be removed without parting the crank case. The water-jackets are of spun copper, brazed on to the cylinder, and corrugated so as to admit of free expansion; the water is circulated by means of a centrifugal pump. The pistons are of cast-iron, each fitted with three rings, and the connecting rods are of high grade steel, machined all over and fitted with bushes of phosphor bronze; these rods are connected to a central collar, carried on the crank pin by two ball-bearings. The crankshaft has a single throw, and is made in two parts to allow the cage for carrying the big end-pins of the connecting rods to be placed in position.

The casing is in two parts, on one of which the brackets for fixing the engine are carried, while the other part carries the valve-gear. Bolts secure the two parts together. The mechanically-operated steel valves on the cylinders are each fitted with double springs and the valves are operated by rods and levers. Two Zenith carburettors are fitted on the rear half of the crank case, and short induction pipes are led to each cylinder; each of the carburettors is heated by the exhaust gases. Ignition is by two high-tension magnetos, and a compressed air self-starting arrangement is provided. Two oil pumps are fitted for lubricating purposes, one of which forces oil to the crankshaft and connecting-rod bearings, while the second forces oil to the valve gear, the cylinders being so arranged that the oil which flows along the walls cannot flood the lower cylinders. This engine operates upon a six-stroke cycle, a rather rare arrangement for internal combustion engines of the electrical ignition type; this is done in order to obtain equal angular intervals for the working impulses imparted to the rotating crankshaft, as the cylinders are arranged in groups of seven, and all act upon the one crankshaft. The angle, therefore, between the impulses is 77 1/7 degrees. A diagram is inset giving a side view of the engine, in order to show the grouping of the cylinders.

The 600 horse-power Salmson engine was designed with a view to fitting to airships, and was in reality two nine-cylindered engines, with a gear-box connecting them; double air-screws were fitted, and these were so arranged that either or both of them might be driven by either or both engines; in addition to this, the two engines were complete and separate engines as regards carburation and ignition, etc., so that they could be run independently of each other. The cylinders were exceptionally 'long stroke,' being 5.9 inches bore to 8.27 inches stroke, and the rated power was developed at 1,200 revolutions per minute, the weight of the complete engine being only 4.1 lbs. per horse-power at the normal rating.

A type of engine specially devised for airship propulsion is that in which the cylinders are arranged horizontally instead of vertically, the main advantages of this form being the reduction of head resistance and less obstruction to the view of the pilot. A casing, mounted on the top of the engine, supports the air-screw, which is driven through bevel gearing from the upper end of the crankshaft. With this type of engine a better rate of air-screw efficiency is obtained by gearing the screw down to half the rate of revolution of the engine, this giving a more even torque. The petrol consumption of the type is very low, being only 0.48 lbs. per horse-power per hour, and equal economy is claimed as regards lubricating oil, a consumption of as little as 0.04 lbs. per horse-power per hour being claimed.

Certain American radial engines were made previous to 1914, the principal being the Albatross six-cylinder engines of 50 and 100 horse-powers. Of these the smaller size was air-cooled, with cylinders of 4.5 inches bore and 5 inches stroke, developing the rated power at 1,230 revolutions per minute, with a weight of about 5 lbs. per horse-power. The 100 horse-power size had cylinders of 5.5 inches bore, developing its rated power at 1,230 revolutions per minute, and weighing only 2.75 lbs. per horse-power. This engine was markedly similar to the six-cylindered Anzani, having all the valves mechanically operated, and with auxiliary exhaust ports at the bottoms of the cylinders, overrun by long pistons. These Albatross engines had their cylinders arranged in two groups of three, with each group of three pistons operating on one of two crank pins, each 180 degrees apart.

The radial type of engine, thanks to Charles Manly, had the honour of being first in the field as regards aero work. Its many advantages, among which may be specially noted the very short crankshaft as compared with vertical, Vee, or 'broad arrow' type of engine, and consequent greater rigidity, ensure it consideration by designers of to-day, and render it certain that the type will endure. Enthusiasts claim that the 'broad arrow' type, or Vee with a third row of cylinders inset between the original two, is just as much a development from the radial engine as from the vertical and resulting Vee; however this may be, there is a place for the radial type in air-work for as long as the internal combustion engine remains as a power plant.

IV. THE ROTARY TYPE

M. Laurent Seguin, the inventor of the Gnome rotary aero engine, provided as great a stimulus to aviation as any that was given anterior to the war period, and brought about a great advance in mechanical flight, since these well-made engines gave a high-power output for their weight, and were extremely smooth in running. In the rotary design the crankshaft of the engine is stationary, and the cylinders, crank case, and all their adherent parts rotate; the working is thus exactly opposite in principle to that of the radial type of aero engine, and the advantage of the rotary lies in the considerable flywheel effect produced by the revolving cylinders, with consequent evenness of torque. Another advantage is that air-cooling, adopted in all the Gnome engines, is rendered much more effective by the rotation of the cylinders, though there is a tendency to distortion through the leading side of each cylinder being more efficiently cooled than the opposite side; advocates of other types are prone to claim that the air resistance to the revolving cylinders absorbs some 10 per cent of the power developed by the rotary engine, but that has not prevented the rotary from attaining to great popularity as a prime mover.

There were, in the list of aero engines compiled in 1910, five rotary engines included, all air-cooled. Three of these were Gnome engines, and two of the make known as 'International.' They ranged from 21.5 to 123 horse-power, the latter being rated at only 1.8 lbs. weight per brake horse-power, and having fourteen cylinders, 4.33 inches in diameter by 4.7 inches stroke. By 1914 forty-three different sizes and types of rotary engine were being constructed, and in 1913 five rotary type engines were entered for the series of aeroplane engine trials held in Germany. Minor defects ruled out four of these, and only the German Bayerischer Motoren Flugzeugwerke completed the seven-hour test prescribed for competing engines. Its large fuel consumption barred this engine from the final trials, the consumption being some 0.95 pints per horse-power per hour. The consumption of lubricating oil, also was excessive, standing at 0.123 pint per horse-power per hour. The engine gave 37.5 effective horse-power during its trial, and the loss due to air resistance was 4.6 horse-power, about 11 per cent. The accompanying drawing shows the construction of the engine, in which the seven cylinders are arranged radially on the crank case; the method of connecting the pistons to the crank pins can be seen. The mixture is drawn through the crank chamber, and to enter the cylinder it passes through the two automatic valves in the crown of the piston; the exhaust valves are situated in the tops of the cylinders, and are actuated by cams and push-rods. Cooling of the cylinder is assisted by the radial rings, and the diameter of these rings is increased round the hottest part of the cylinder. When long flights are undertaken the advantage of the light weight of this engine is more than counterbalanced by its high fuel and lubricating oil consumption, but there are other makes which are much better than this seven-cylinder German in respect of this.

Rotation of the cylinders in engines of this type is produced by the side pressure of the pistons on the cylinder walls, and in order to prevent this pressure from becoming abnormally large it is necessary to keep the weight of the piston as low as possible, as the pressure is produced by the tangential acceleration and retardation of the piston. On the upward stroke the circumferential velocity of the piston is rapidly increased, which causes it to exert a considerable tangential pressure on the side of the cylinder, and on the return stroke there is a corresponding retarding effect due to the reduction of the circumferential velocity of the piston. These side pressures cause an appreciable increase in the temperatures of the cylinders and pistons, which makes it necessary to keep the power rating of the engines fairly low.

Seguin designed his first Gnome rotary as a 34 horse-power engine when run at a speed of 1,300 revolutions per minute. It had five cylinders, and the weight was 3.9 lbs. per horse-power. A seven-cylinder model soon displaced this first engine, and this latter, with a total weight of 165 lbs., gave 61.5 horse-power. The cylinders were machined out of solid nickel chrome-steel ingots, and the machining was carried out so that the cylinder walls were under 1/6 of an inch in thickness. The pistons were cast-iron, fitted each with two rings, and the automatic inlet valve to the cylinder was placed in the crown of the piston. The connecting rods, of 'H' section, were of nickel chrome-steel, and the large end of one rod, known as the 'master-rod' embraced the crank pin; on the end of this rod six hollow steel pins were carried, and to these the remaining six connecting-rods were attached. The crankshaft of the engine was made of nickel chrome-steel, and was in two parts connected together at the crank pin; these two parts, after the master-rod had been placed in position and the other connecting rods had been attached to it, were firmly secured. The steel crank case was made in five parts, the two central ones holding the cylinders in place, and on one side another of the five castings formed a cam-box, to the outside of which was secured the extension to which the air-screw was attached. On the other side of the crank case another casting carried the thrust-box, and the whole crank case, with its cylinders and gear, was carried on the fixed crank shaft by means of four ball-bearings, one of which also took the axial thrust of the air-screw.

For these engines, castor oil is the lubricant usually adopted, and it is pumped to the crankshaft by means of a gear-driven oil pump; from this shaft the other parts of the engine are lubricated by means of centrifugal force, and in actual practice sufficient unburnt oil passes through the cylinders to lubricate the exhaust valve, which partly accounts for the high rate of consumption of lubricating oil. A very simple carburettor of the float less, single-spray type was used, and the mixture was passed along the hollow crankshaft to the interior of the crank case, thence through the automatic inlet valves in the tops of the pistons to the combustion chambers of the cylinders. Ignition was by means of a high-tension magneto specially geared to give the correct timing, and the working impulses occurred at equal angular intervals of 102.85 degrees. The ignition was timed so that the firing spark occurred when the cylinder was 26 degrees before the position in which the piston was at the outer end of its stroke, and this timing gave a maximum pressure in the cylinder just after the piston had passed this position.

By 1913, eight different sizes of the Gnome engine were being constructed, ranging from 45 to 180 brake horse-power; four of these were single-crank engines one having nine and the other three having seven cylinders. The remaining four were constructed with two cranks; three of them had fourteen cylinders apiece, ranged in groups of seven, acting on the cranks, and the one other had eighteen cylinders ranged in two groups of nine, acting on its two cranks. Cylinders of the two-crank engines are so arranged (in the fourteen-cylinder type) that fourteen equal angular impulses occur during each cycle; these engines are supported on bearings on both sides of the engine, the air-screw being placed outside the front support. In the eighteen-cylinder model the impulses occur at each 40 degrees of angular rotation of the cylinders, securing an extremely even rotation of the air-screw.

In 1913 the Gnome Monosoupape engine was introduced, a model in which the inlet valve to the cylinder was omitted, while the piston was of the ordinary cast-iron type. A single exhaust valve in the cylinder head was operated in a manner similar to that on the previous Gnome engines, and the fact of this being the only valve on the cylinder gave the engine its name. Each cylinder contained ports at the bottom which communicated with the crank chamber, and were overrun by the piston when this was approaching the bottom end of its stroke. During the working cycle of the engine the exhaust valve was opened early to allow the exhaust gases to escape from the cylinder, so that by the time the piston overran the ports at the bottom the pressure within the cylinder was approximately equal to that in the crank case, and practically no flow of gas took place in either direction through the ports. The exhaust valve remained open as usual during the succeeding up-stroke of the piston, and the valve was held open until the piston had returned through about one-third of its downward stroke, thus permitting fresh air to enter the cylinder. The exhaust valve then closed, and the downward motion of the piston, continuing, caused a partial vacuum inside the cylinder; when the piston overran the ports, the rich mixture from the crank case immediately entered. The cylinder was then full of the mixture, and the next upward stroke of the piston compressed the charge; upon ignition the working cycle was repeated. The speed variation of this engine was obtained by varying the extent and duration of the opening of the exhaust valves, and was controlled by the pilot by hand-operated levers acting on the valve tappet rollers. The weight per horsepower of these engines was slightly less than that of the two-valve type, while the lubrication of the gudgeon pin and piston showed an improvement, so that a lower lubricating oil consumption was obtained. The 100 horse-power Gnome Monosoupape was built with nine cylinders, each 4.33 inches bore by 5.9 inches stroke, and it developed its rated power at 1,200 revolutions per minute.

An engine of the rotary type, almost as well known as the Gnome, is the Clerget, in which both cylinders and crank case are made of steel, the former having the usual radial fins for cooling. In this type the inlet and exhaust valves are both located in the cylinder head, and mechanically operated by push-rods and rockers. Pipes are carried from the crank case to the inlet valve casings to convey the mixture to the cylinders, a carburettor of the central needle type being used. The carburetted mixture is taken into the crank case chamber in a manner similar to that of the Gnome engine. Pistons of aluminium alloy, with three cast-iron rings, are fitted, the top ring being of the obturator type. The large end of one of the nine connecting rods embraces the crank pin and the pressure is taken on two ball-bearings housed in the end of the rod. This carries eight pins, to which the other rods are attached, and the main rod being rigid between the crank pin and piston pin determines the position of the pistons. Hollow connecting-rods are used, and the lubricating oil for the piston pins passes from the crankshaft through the centres of the rods. Inlet and exhaust valves can be set quite independently of one another--a useful point, since the correct timing of the opening of these valves is of importance. The inlet valve opens 4 degrees from top centre and closes after the bottom dead centre of the piston; the exhaust valve opens 68 degrees before the bottom centre and closes 4 degrees after the top dead centre of the piston. The magnetos are set to give the spark in the cylinder at 25 degrees before the end of the compression stroke--two high-tension magnetos are used: if desired, the second one can be adjusted to give a later spark for assisting the starting of the engine. The lubricating oil pump is of the valveless two-plunger type, so geared that it runs at seven revolutions to 100 revolutions of the engine; by counting the pulsations the speed of the engine can be quickly calculated by multiplying the pulsations by 100 and dividing by seven. In the 115 horse-power nine-cylinder Clerget the cylinders are 4.7 bore with a 6.3 inches stroke, and the rated power of the engine is obtained at 1,200 revolutions per minute. The petrol consumption is 0.75 pint per horse-power per hour.

A third rotary aero engine, equally well known with the foregoing two, is the Le Rhone, made in four different sizes with power outputs of from 50 to 160 horse-power; the two smaller sizes are single crank engines with seven and nine cylinders respectively, and the larger sizes are of double-crank design, being merely the two smaller sizes doubled--fourteen and eighteen-cylinder engines. The inlet and exhaust valves are located in the cylinder head, and both valves are mechanically operated by one push-rod and rocker, radial pipes from crank case to inlet valve casing taking the mixture to the cylinders. The exhaust valves are placed on the leading, or air-screw side, of the engine, in order to get the fullest possible cooling effect. The rated power of each type of engine is obtained at 1,200 revolutions per minute, and for all four sizes the cylinder bore is 4.13 inches, with a 5.5 inches piston stroke. Thin cast-iron liners are shrunk into the steel cylinders in order to reduce the amount of piston friction. Although the Le Rhone engines are constructed practically throughout of steel, the weight is only 2.9 lbs. per horse-power in the eighteen-cylinder type.

American enterprise in the construction of the rotary type is perhaps best illustrated in the 'Gyro 'engine; this was first constructed with inlet valves in the heads of the pistons, after the Gnome pattern, the exhaust valves being in the heads of the cylinders. The inlet valve in the crown of each piston was mechanically operated in a very ingenious manner by the oscillation of the connecting-rod. The Gyro-Duplex engine superseded this original design, and a small cross-section illustration of this is appended. It is constructed in seven and nine-cylinder sizes, with a power range of from 50 to 100 horse-power; with the largest size the low weight of 2.5 lbs.. per horse-power is reached. The design is of considerable interest to the internal combustion engineer, for it embodies a piston valve for controlling auxiliary exhaust ports, which also acts as the inlet valve to the cylinder. The piston uncovers the auxiliary ports when it reaches the bottom of its stroke, and at the end of the power stroke the piston is in such a position that the exhaust can escape over the top of it. The exhaust valve in the cylinder head is then opened by means of the push-rod and rocker, and is held open until the piston has completed its upward stroke and returned through more than half its subsequent return stroke. When the exhaust valve closes, the cylinder has a charge of fresh air, drawn in through the exhaust valve, and the further motion of the piston causes a partial vacuum; by the time the piston reaches bottom dead centre the piston-valve has moved up to give communication between the cylinder and the crank case, therefore the mixture is drawn into the cylinder. Both the piston valve and exhaust valve are operated by cams formed on the one casting, which rotates at seven-eighths engine speed for the seven-cylinder type, and nine-tenths engine speed for the nine-cylinder engines. Each of these cams has four or five points respectively, to suit the number of cylinders.

The steel cylinders are machined from solid forgings and provided with webs for air-cooling as shown. Cast-iron pistons are used, and are connected to the crankshaft in the same manner as with the Gnome and Le Rhone engines. Petrol is sprayed into the crank case by a small geared pump and the mixture is taken from there to the piston valves by radial pipes. Two separate pumps are used for lubrication, one forcing oil to the crank-pin bearing and the other spraying the cylinders.

Among other designs of rotary aero engines the E.J.C. is noteworthy, in that the cylinders and crank case of this engine rotate in opposite directions, and two air-screws are used, one being attached to the end of the crankshaft, and the other to the crank case. Another interesting type is the Burlat rotary, in which both the cylinders and crankshaft rotate in the same direction, the rotation of the crankshaft being twice that of the cylinders as regards speed. This engine is arranged to work on the four-stroke cycle with the crankshaft making four, and the cylinders two, revolutions per cycle.

It would appear that the rotary type of engine is capable of but little more improvement--save for such devices as these of the last two engines mentioned, there is little that Laurent Seguin has not already done in the Gnome type. The limitation of the rotary lies in its high fuel and lubricating oil consumption, which renders it unsuited for long-distance aero work; it was, in the war period, an admirable engine for such short runs as might be involved in patrol work 'over the lines,' and for similar purposes, but the watercooled Vee or even vertical, with its much lower fuel consumption, was and is to be preferred for distance work. The rotary air-cooled type has its uses, and for them it will probably remain among the range of current types for some time to come. Experience of matters aeronautical is sufficient to show, however, that prophecy in any direction is most unsafe.

V. THE HORIZONTALLY-OPPOSED ENGINE

Among the first internal combustion engines to be taken into use with aircraft were those of the horizontally-opposed four-stroke cycle type, and, in every case in which these engines were used, their excellent balance and extremely even torque rendered them ideal-until the tremendous increase in power requirements rendered the type too long and bulky for placing in the fuselage of an aeroplane. As power increased, there came a tendency toward placing cylinders radially round a central crankshaft, and, as in the case of the early Anzani, it may be said that the radial engine grew out of the horizontal opposed piston type. There were, in 1910--that is, in the early days of small power units, ten different sizes of the horizontally opposed engine listed for manufacture, but increase in power requirements practically ruled out the type for air work.

The Darracq firm were the leading makers of these engines in 1910; their smallest size was a 24 horsepower engine, with two cylinders each of 5.1 inches bore by 4.7 inches stroke. This engine developed its rated power at 1,500 revolutions per minute, and worked out at a weight of 5 lbs. per horse-power. With these engines the cranks are so placed that two regular impulses are given to the crankshaft for each cycle of working, an arrangement which permits of very even balancing of the inertia forces of the engine. The Darracq firm also made a four-cylindered horizontal opposed piston engine, in which two revolutions were given to the crankshaft per revolution, at equal angular intervals.

The Dutheil-Chambers was another engine of this type, and had the distinction of being the second largest constructed. At 1,000 revolutions per minute it developed 97 horse-power; its four cylinders were each of 4.93 inches bore by 11.8 inches stroke--an abnormally long stroke in comparison with the bore. The weight--which owing to the build of the engine and its length of stroke was bound to be rather high, actually amounted to 8.2 lbs. per horse-power. Water cooling was adopted, and the engine was, like the Darracq four-cylinder type, so arranged as to give two impulses per revolution at equal angular intervals of crankshaft rotation.

One of the first engines of this type to be constructed in England was the Alvaston, a water-cooled model which was made in 20, 30, and 50 brake horse-power sizes, the largest being a four-cylinder engine. All three sizes were constructed to run at 1,200 revolutions per minute. In this make the cylinders were secured to the crank case by means of four long tie bolts passing through bridge pieces arranged across the cylinder heads, thus relieving the cylinder walls of all longitudinal explosion stresses. These bridge pieces were formed from chrome vanadium steel and milled to an 'H' section, and the bearings for the valve-tappet were forged solid with them. Special attention was given to the machining of the interiors of the cylinders and the combustion heads, with the result that the exceptionally high compression of 95 lbs. per square inch was obtained, giving a very flexible engine. The cylinder heads were completely water-jacketed, and copper water-jackets were also fitted round the cylinders. The mechanically operated valves were actuated by specially shaped cams, and were so arranged that only two cams were required for the set of eight valves. The inlet valves at both ends of the engine were connected by a single feed-pipe to which the carburettor was attached, the induction piping being arranged above the engine in an easily accessible position. Auxiliary air ports were provided in the cylinder walls so that the pistons overran them at the end of their stroke. A single vertical shaft running in ball-bearings operated the valves and water circulating pump, being driven by spiral gearing from the crankshaft at half speed. In addition to the excellent balance obtained with this engine, the makers claimed with justice that the number of working parts was reduced to an absolute minimum.

In the two-cylinder Darracq, the steel cylinders were machined from solid, and auxiliary exhaust ports, overrun by the piston at the inner end of its stroke, were provided in the cylinder walls, consisting of a circular row of drilled holes--this arrangement was subsequently adopted on some of the Darracq racing car engines. The water jackets were of copper, soldered to the cylinder walls; both the inlet and exhaust valves were located in the cylinder heads, being operated by rockers and push-rods actuated by cams on the halftime shaft driven from one end of the crankshaft. Ignition was by means of a high-tension magneto, and long induction pipes connected the-ends of the cylinders to the carburettor, the latter being placed underneath the engine. Lubrication was effected by spraying oil into the crank case by means of a pump, and a second pump circulated the cooling water.

Another good example of this type of engine was the Eole, which had eight opposed pistons, each pair of which was actuated by a common combustion chamber at the centre of the engine, two crankshafts being placed at the outer ends of the engine. This reversal of the ordinary arrangement had two advantages; it simplified induction, and further obviated the need for cylinder heads, since the explosion drove at two piston heads instead of at one piston head and the top of the cylinder; against this, however, the engine had to be constructed strongly enough to withstand the longitudinal stresses due to the explosions, as the cranks are placed on the outer ends and the cylinders and crank-cases take the full force of each explosion. Each crankshaft drove a separate air-screw.

This pattern of engine was taken up by the Dutheil-Chambers firm in the pioneer days of aircraft, when the firm in question produced seven different sizes of horizontal engines. The Demoiselle monoplane used by Santos-Dumont in 1909 was fitted with a two-cylinder, horizontally-opposed Dutheil-Chambers engine, which developed 25 brake horse-power at a speed of 1,100 revolutions per minute, the cylinders being of 5 inches bore by 5.1 inches stroke, and the total weight of the engine being some 120 lbs. The crankshafts of these engines were usually fitted with steel flywheels in order to give a very even torque, the wheels being specially constructed with wire spokes. In all the Dutheil-Chambers engines water cooling was adopted, and the cylinders were attached to the crank cases by means of long bolts passing through the combustion heads.

For their earliest machines, the Clement-Bayard firm constructed horizontal engines of the opposed piston type. The best known of these was the 30 horse-power size, which had cylinders of 4.7 inches diameter by 5.1 inches stroke, and gave its rated power at 1,200 revolutions per minute. In this engine the steel cylinders were secured to the crank case by flanges, and radiating ribs were formed around the barrel to assist the air-cooling. Inlet and exhaust valves were actuated by push-rods and rockers actuated from the second motion shaft mounted above the crank case; this shaft also drove the high-tension magneto with which the engine was fitted. A ring of holes drilled round each cylinder constituted auxiliary ports which the piston uncovered at the inner end of its stroke, and these were of considerable assistance not only in expelling exhaust gases, but also in moderating the temperature of the cylinder and of the main exhaust valve fitted in the cylinder head. A water-cooled Clement-Bayard horizontal engine was also made, and in this the auxiliary exhaust ports were not embodied; except in this particular, the engine was very similar to the water-cooled Darracq.

The American Ashmusen horizontal engine, developing 100 horse-power, is probably the largest example of this type constructed. It was made with six cylinders arranged on each side of a common crank case, with long bolts passing through the cylinder heads to assist in holding them down. The induction piping and valve-operating gear were arranged below the engine, and the half-speed shaft carried the air-screw.

Messrs Palons and Beuse, Germans, constructed a light-weight, air-cooled, horizontally-opposed engine, two-cylindered. In this the cast-iron cylinders were made very thin, and were secured to the crank case by bolts passing through lugs cast on the outer ends of the cylinders; the crankshaft was made hollow, and holes were drilled through the webs of the connecting-rods in order to reduce the weight. The valves were fitted to the cylinder heads, the inlet valves being of the automatic type, while the exhaust valves were mechanically operated from the cam-shaft by means of rockers and push-rods. Two carburettors were fitted, to reduce the induction piping to a minimum; one was attached to each combustion chamber, and ignition was by the normal high-tension magneto driven from the halftime shaft.

There was also a Nieuport two-cylinder air-cooled horizontal engine, developing 35 horse-power when running at 1,300 revolutions per minute, and being built at a weight of 5.1 lbs. per horse-power. The cylinders were of 5.3 inches diameter by 5.9 inches stroke; the engine followed the lines of the Darracq and Dutheil-Chambers pretty closely, and thus calls for no special description.

The French Kolb-Danvin engine of the horizontal type, first constructed in 1905, was probably the first two-stroke cycle engine designed to be applied to the propulsion of aircraft; it never got beyond the experimental stage, although its trials gave very good results. Stepped pistons were adopted, and the charging pump at one end was used to scavenge the power cylinder at the other ends of the engine, the transfer ports being formed in the main casting. The openings of these ports were controlled at both ends by the pistons, and the location of the ports appears to have made it necessary to take the exhaust from the bottom of one cylinder and from the top of the other. The carburetted mixture was drawn into the scavenging cylinders, and the usual deflectors were cast on the piston heads to assist in the scavenging and to prevent the fresh gas from passing out of the exhaust ports.

VI. THE TWO-STROKE CYCLE ENGINE

Although it has been little used for aircraft propulsion, the possibilities of the two-stroke cycle engine render some study of it desirable in this brief review of the various types of internal combustion engine applicable both to aeroplanes and airships. Theoretically the two-stroke cycle engine--or as it is more commonly termed, the 'two-stroke,' is the ideal power producer; the doubling of impulses per revolution of the crankshaft should render it of very much more even torque than the four-stroke cycle types, while, theoretically, there should be a considerable saving of fuel, owing to the doubling of the number of power strokes per total of piston strokes. In practice, however, the inefficient scavenging of virtually every two-stroke cycle engine produced nullifies or more than nullifies its advantages over the four-stroke cycle engine; in many types, too, there is a waste of fuel gases through the exhaust ports, and much has yet to be done in the way of experiment and resulting design before the two-stroke cycle engine can be regarded as equally reliable, economical, and powerful with its elder brother.

The first commercially successful engine operating on the two-stroke cycle was invented by Mr Dugald Clerk, who in 1881 proved the design feasible. As is more or less generally understood, the exhaust gases of this engine are discharged from the cylinder during the time that the piston is passing the inner dead centre, and the compression, combustion, and expansion of the charge take place in similar manner to that of the four-stroke cycle engine. The exhaust period is usually controlled by the piston overrunning ports in the cylinder at the end of its working stroke, these ports communicating direct with the outer air--the complication of an exhaust valve is thus obviated; immediately after the escape of the exhaust gases, charging of the cylinder occurs, and the fresh gas may be introduced either through a valve in the cylinder head or through ports situated diametrically opposite to the exhaust ports. The continuation of the outward stroke of the piston, after the exhaust ports have been closed, compresses the charge into the combustion chamber of the cylinder, and the ignition of the mixture produces a recurrence of the working stroke.

Thus, theoretically, is obtained the maximum of energy with the minimum of expenditure; in practice, however, the scavenging of the power cylinder, a matter of great importance in all internal combustion engines, is often imperfect, owing to the opening of the exhaust ports being of relatively short duration; clearing the exhaust gases out of the cylinder is not fully accomplished, and these gases mix with the fresh charge and detract from its efficiency. Similarly, owing to the shorter space of time allowed, the charging of the cylinder with the fresh mixture is not so efficient as in the four-stroke cycle type; the fresh charge is usually compressed slightly in a separate chamber--crank case, independent cylinder, or charging pump, and is delivered to the working cylinder during the beginning of the return stroke of the piston, while in engines working on the four-stroke cycle principle a complete stroke is devoted to the expulsion of the waste gases of the exhaust, and another full stroke to recharging the cylinder with fresh explosive mixture.

Theoretically the two-stroke and the four-stroke cycle engines possess exactly the same thermal efficiency, but actually this is modified by a series of practical conditions which to some extent tend to neutralise the very strong case in favour of the two-stroke cycle engine. The specific capacity of the engine operating on the two-stroke principle is theoretically twice that of one operating on the four-stroke cycle, and consequently, for equal power, the former should require only about half the cylinder volume of the latter; and, owing to the greater superficial area of the smaller cylinder, relatively, the latter should be far more easily cooled than the larger four-stroke cycle cylinder; thus it should be possible to get higher compression pressures, which in turn should result in great economy of working. Also the obtaining of a working impulse in the cylinder for each revolution of the crankshaft should give a great advantage in regularity of rotation--which it undoubtedly does--and the elimination of the operating gear for the valves, inlet and exhaust, should give greater simplicity of design.

In spite of all these theoretical--and some practical--advantages the four-stroke cycle engine was universally adopted for aircraft work; owing to the practical equality of the two principles of operation, so far as thermal efficiency and friction losses are concerned, there is no doubt that the simplicity of design (in theory) and high power output to weight ratio (also in theory) ought to have given the 'two-stroke' a place on the aeroplane. But this engine has to be developed so as to overcome its inherent drawbacks; better scavenging methods have yet to be devised--for this is the principal drawback--before the two-stroke can come to its own as a prime mover for aircraft.

Mr Dugald Clerk's original two-stroke cycle engine is indicated roughly, as regards principle, by the accompanying diagram, from which it will be seen that the elimination of the ordinary inlet and exhaust valves of the four-stroke type is more than compensated by a separate cylinder which, having a piston worked from the connecting-rod of the power cylinder, was used to charging, drawing the mixture from the carburettor past the valve in the top of the charging cylinder, and then forcing it through the connecting pipe into the power cylinder. The inlet valves both on the charging and the power cylinders are automatic; when the power piston is near the bottom of its stroke the piston in the charging cylinder is compressing the carburetted air, so that as soon as the pressure within the power cylinder is relieved by the exit of the burnt gases through the exhaust ports the pressure in the charging cylinder causes the valve in the head of the power cylinder to open, and fresh mixture flows into the cylinder, replacing the exhaust gases. After the piston has again covered the exhaust ports the mixture begins to be compressed, thus automatically closing the inlet valve. Ignition occurs near the end of the compression stroke, and the working stroke immediately follows, thus giving an impulse to the crankshaft on every down stroke of the piston. If the scavenging of the cylinder were complete, and the cylinder were to receive a full charge of fresh mixture for every stroke, the same mean effective pressure as is obtained with four-stroke cycle engines ought to be realised, and at an equal speed of rotation this engine should give twice the power obtainable from a four-stroke cycle engine of equal dimensions. This result was not achieved, and, with the improvements in construction brought about by experiment up to 1912, the output was found to be only about fifty per cent more than that of a four-stroke cycle engine of the same size, so that, when the charging cylinder is included, this engine has a greater weight per horsepower, while the lowest rate of fuel consumption recorded was 0.68 lb. per horsepower per hour.

In 1891 Mr Day invented a two-stroke cycle engine which used the crank case as a scavenging chamber, and a very large number of these engines have been built for industrial purposes. The charge of carburetted air is drawn through a non-return valve into the crank chamber during the upstroke of the piston, and compressed to about 4 lbs. pressure per square inch on the down stroke. When the piston approaches the bottom end of its stroke the upper edge first overruns an exhaust port, and almost immediately after uncovers an inlet port on the opposite side of the cylinder and in communication with the crank chamber; the entering charge, being under pressure, assists in expelling the exhaust gases from the cylinder. On the next upstroke the charge is compressed into the combustion space of the cylinder, a further charge simultaneously entering the crank case to be compressed after the ignition for the working stroke. To prevent the incoming charge escaping through the exhaust ports of the cylinder a deflector is formed on the top of the piston, causing the fresh gas to travel in an upward direction, thus avoiding as far as possible escape of the mixture to the atmosphere. From experiments conducted in 1910 by Professor Watson and Mr Fleming it was found that the proportion of fresh gases which escaped unburnt through the exhaust ports diminished with increase of speed; at 600 revolutions per minute about 36 per cent of the fresh charge was lost; at 1,200 revolutions per minute this was reduced to 20 per cent, and at 1,500 revolutions it was still farther reduced to 6 per cent.

So much for the early designs. With regard to engines of this type specially constructed for use with aircraft, three designs call for special mention. Messrs A. Gobe and H. Diard, Parisian engineers, produced an eight-cylindered two-stroke cycle engine of rotary design, the cylinders being co-axial. Each pair of opposite pistons was secured together by a rigid connecting rod, connected to a pin on a rotating crankshaft which was mounted eccentrically to the axis of rotation of the cylinders. The crankshaft carried a pinion gearing with an internally toothed wheel on the transmission shaft which carried the air-screw. The combustible mixture, emanating from a common supply pipe, was led through conduits to the front ends of the cylinders, in which the charges were compressed before being transferred to the working spaces through ports in tubular extensions carried by the pistons. These extensions had also exhaust ports, registering with ports in the cylinder which communicated with the outer air, and the extensions slid over depending cylinder heads attached to the crank case by long studs. The pump charge was compressed in one end of each cylinder, and the pump spaces each delivered into their corresponding adjacent combustion spaces. The charges entered the pump spaces during the suction period through passages which communicated with a central stationary supply passage at one end of the crank case, communication being cut off when the inlet orifice to the passage passed out of register with the port in the stationary member. The exhaust ports at the outer end of the combustion space opened just before and closed a little later than the air ports, and the incoming charge assisted in expelling the exhaust gases in a manner similar to that of the earlier types of two-stroke cycle engine; The accompanying rough diagram assists in showing the working of this engine.

Exhibited in the Paris Aero Exhibition of 1912, the Laviator two-stroke cycle engine, six-cylindered, could be operated either as a radial or as a rotary engine, all its pistons acting on a single crank. Cylinder dimensions of this engine were 3.94 inches bore by 5.12 inches stroke, and a power output of 50 horse-power was obtained when working at a rate of 1,200 revolutions per minute. Used as a radial engine, it developed 65 horse-power at the same rate of revolution, and, as the total weight was about 198 lbs., the weight of about 3 lbs. per horse-power was attained in radial use. Stepped pistons were employed, the annular space between the smaller or power piston and the walls of the larger cylinder being used as a charging pump for the power cylinder situated 120 degrees in rear of it. The charging cylinders were connected by short pipes to ports in the crank case which communicated with the hollow crankshaft through which the fresh gas was supplied, and once in each revolution each port in the case registered with the port in the hollow shaft. The mixture which then entered the charging cylinder was transferred to the corresponding working cylinder when the piston of that cylinder had reached the end of its power stroke, and immediately before this the exhaust ports diametrically opposite the inlet ports were uncovered; scavenging was thus assisted in the usual way. The very desirable feature of being entirely valveless was accomplished with this engine, which is also noteworthy for exceedingly compact design.

The Lamplough six-cylinder two-stroke cycle rotary, shown at the Aero Exhibition at Olympia in 1911, had several innovations, including a charging pump of rotary blower type. With the six cylinders, six power impulses at regular intervals were given on each rotation; otherwise, the cycle of operations was carried out much as in other two-stroke cycle engines. The pump supplied the mixture under slight pressure to an inlet port in each cylinder, which was opened at the same time as the exhaust port, the period of opening being controlled by the piston. The rotary blower sucked the mixture from the carburettor and delivered it to a passage communicating with the inlet ports in the cylinder walls. A mechanically-operated exhaust valve was placed in the centre of each cylinder head, and towards the end of the working stroke this valve opened, allowing part of the burnt gases to escape to the atmosphere; the remainder was pushed out by the fresh mixture going in through the ports at the bottom end of the cylinder. In practice, one or other of the cylinders was always taking fresh mixture while working, therefore the delivery from the pump was continuous and the mixture had not to be stored under pressure.

The piston of this engine was long enough to keep the ports covered when it was at the top of the stroke, and a bottom ring was provided to prevent the mixture from entering the crank case. In addition to preventing leakage, this ring no doubt prevented an excess of oil working up the piston into the cylinder. As the cylinder fired with every revolution, the valve gear was of the simplest construction, a fixed cam lifting each valve as the cylinder came into position. The spring of the exhaust valve was not placed round the stem in the usual way, but at the end of a short lever, away from the heat of the exhaust gases. The cylinders were of cast steel, the crank case of aluminium, and ball-bearings were fitted to the crankshaft, crank pins, and the rotary blower pump. Ignition was by means of a high-tension magneto of the two-spark pattern, and with a total weight of 300 lbs. the maximum output was 102 brake horse-power, giving a weight of just under 3 lbs. per horse-power.

One of the most successful of the two-stroke cycle engines was that designed by Mr G. F. Mort and constructed by the New Engine Company. With four cylinders of 3.69 inches bore by 4.5 inches stroke, and running at 1,250 revolutions per minute, this engine developed 50 brake horse-power; the total weight of the engine was 155 lbs., thus giving a weight of 3.1 lbs. per horse-power. A scavenging pump of the rotary type was employed, driven by means of gearing from the engine crankshaft, and in order to reduce weight to a minimum the vanes were of aluminium. This engine was tried on a biplane, and gave very satisfactory results.

American design yields two apparently successful two-stroke cycle aero engines. A rotary called the Fredericson engine was said to give an output of 70 brake horse-power with five cylinders 4.5 inches diameter by 4.75 inches stroke, running at 1,000 revolutions per minute. Another, the Roberts two-stroke cycle engine, yielded 100 brake horse-power from six cylinders of the stepped piston design; two carburettors, each supplying three cylinders, were fitted to this engine. Ignition was by means of the usual high-tension magneto, gear-driven from the crankshaft, and the engine, which was water-cooled, was of compact design.

It may thus be seen that the two-stroke cycle type got as far as actual experiment in air work, and that with considerable success. So far, however, the greater reliability of the four-stroke cycle has rendered it practically the only aircraft engine, and the two-stroke has yet some way to travel before it becomes a formidable competitor, in spite of its admitted theoretical and questioned practical advantages.

VII. ENGINES OF THE WAR PERIOD

The principal engines of British, French, and American design used in the war period and since are briefly described under the four distinct types of aero engine; such notable examples as the Rolls-Royce, Sunbeam, and Napier engines have been given special mention, as they embodied--and still embody--all that is best in aero engine practice. So far, however, little has been said about the development of German aero engine design, apart from the early Daimler and other pioneer makes.

At the outbreak of hostilities in 1914, thanks to subsidies to contractors and prizes to aircraft pilots, the German aeroplane industry was in a comparatively flourishing condition. There were about twenty-two establishments making different types of heavier-than-air machines, monoplane and biplane, engined for the most part with the four-cylinder Argus or the six-cylinder Mercedes vertical type engines, each of these being of 100 horse-power--it was not till war brought increasing demands on aircraft that the limit of power began to rise. Contemporary with the Argus and Mercedes were the Austro-Daimler, Benz, and N.A.G., in vertical design, while as far as rotary types were concerned there were two, the Oberursel and the Stahlhertz; of these the former was by far the most promising, and it came to virtual monopoly of the rotary-engined plane as soon as the war demand began. It was practically a copy of the famous Gnome rotary, and thus deserves little description.

Germany, from the outbreak of war, practically, concentrated on the development of the Mercedes engine; and it is noteworthy that, with one exception, increase of power corresponding with the increased demand for power was attained without increasing the number of cylinders. The various models ranged between 75 and 260 horse-power, the latter being the most recent production of this type. The exception to the rule was the eight-cylinder 240 horse-power, which was replaced by the 260 horse-power six-cylinder model, the latter being more reliable and but very slightly heavier. Of the other engines, the 120 horsepower Argus and the 160 and 225 horse-power Benz were the most used, the Oberursel being very largely discarded after the Fokker monoplane had had its day, and the N.A.G. and Austro-Daimler Daimler also falling to comparative disuse. It may be said that the development of the Mercedes engine contributed very largely to such success as was achieved in the war period by German aircraft, and, in developing the engine, the builders were careful to make alterations in such a way as to effect the least possible change in the design of aeroplane to which they were to be fitted. Thus the engine base of the 175 horse-power model coincided precisely with that of the 150 horse-power model, and the 200 and 240 horse-power models retained the same base dimensions. It was estimated, in 1918, that well over eighty per cent of German aircraft was engined with the Mercedes type.

In design and construction, there was nothing abnormal about the Mercedes engine, the keynote throughout being extreme reliability and such simplification of design as would permit of mass production in different factories. Even before the war, the long list of records set up by this engine formed practical application of the wisdom of this policy; Bohn's flight of 24 hours 10 minutes, accomplished on July 10th and 11th, 1914, 9is an instance of this--the flight was accomplished on an Albatross biplane with a 75 horsepower Mercedes engine. The radial type, instanced in other countries by the Salmson and Anzani makes, was not developed in Germany; two radial engines were made in that country before the war, but the Germans seemed to lose faith in the type under war conditions, or it may have been that insistence on standardisation ruled out all but the proved examples of engine.

Details of one of the middle sizes of Mercedes motor, the 176 horse-power type, apply very generally to the whole range; this size was in use up to and beyond the conclusion of hostilities, and it may still be regarded as characteristic of modern (1920) German practice. The engine is of the fixed vertical type, has six cylinders in line, not off-set, and is water-cooled. The cam shaft is carried in a special bronze casing, seated on the immediate top of the cylinders, and a vertical shaft is interposed between crankshaft and camshaft, the latter being driven by bevel gearing.

On this vertical connecting-shaft the water pump is located, serving to steady the motion of the shaft. Extending immediately below the camshaft is another vertical shaft, driven by bevel gears from the crank-shaft, and terminating in a worm which drives the multiple piston oil pumps.

The cylinders are made from steel forgings, as are the valve chamber elbows, which are machined all over and welded together. A jacket of light steel is welded over the valve elbows and attached to a flange on the cylinders, forming a water-cooling space with a section of about 7/16 of an inch. The cylinder bore is 5.5 inches, and the stroke 6.29 inches. The cylinders are attached to the crank case by means of dogs and long through bolts, which have shoulders near their lower ends and are bolted to the lower half of the crank chamber. A very light and rigid structure is thus obtained, and the method of construction won the flattery of imitation by makers of other nationality.

The cooling system for the cylinders is extremely efficient. After leaving the water pump, the water enters the top of the front cylinders and passes successively through each of the six cylinders of the row; short tubes, welded to the tops of the cylinders, serve as connecting links in the system. The Panhard car engines for years were fitted with a similar cooling system, and the White and Poppe lorry engines were also similarly fitted; the system gives excellent cooling effect where it is most needed, round the valve chambers and the cylinder heads.

The pistons are built up from two pieces; a dropped forged steel piston head, from which depend the piston pin bosses, is combined with a cast-iron skirt, into which the steel head is screwed. Four rings are fitted, three at the upper and one at the lower end of the piston skirt, and two lubricating oil grooves are cut in the skirt, in addition to the ring grooves. Two small rivets retain the steel head on the piston skirt after it has been screwed into position, and it is also welded at two points. The coefficient of friction between the cast-iron and steel is considerably less than that which would exist between two steel parts, and there is less tendency for the skirt to score the cylinder walls than would be the case if all steel were used--so noticeable is this that many makers, after giving steel pistons a trial, discarded them in favour of cast-iron; the Gnome is an example of this, being originally fitted with a steel piston carrying a brass ring, discarded in favour of a cast-iron piston with a percentage of steel in the metal mixture. In the Le Rhone engine the difficulty is overcome by a cast-iron liner to the cylinders.

The piston pin of the Mercedes is of chrome nickel steel, and is retained in the piston by means of a set screw and cotter pin. The connecting rods, of I section, are very short and rigid, carrying floating bronze bushes which fit the piston pins at the small end, and carrying an oil tube on each for conveying oil from the crank pin to the piston pin.

The crankshaft is of chrome nickel steel, carried on seven bearings. Holes are drilled through each of the crank pins and main bearings, for half the diameter of the shaft, and these are plugged with pressed brass studs. Small holes, drilled through the crank cheeks, serve to convey lubricant from the main bearings to the crank pins. The propeller thrust is taken by a simple ball thrust bearing at the propeller end of the crankshaft, this thrust bearing being seated in a steel retainer which is clamped between the two halves of the crank case. At the forward end of the crankshaft there is mounted a master bevel gear on six splines; this bevel floats on the splines against a ball thrust bearing, and, in turn, the thrust is taken by the crank case cover. A stuffing box prevents the loss of lubricant out of the front end of the crank chamber, and an oil thrower ring serves a similar purpose at the propeller end of the crank chamber.

With a motor speed of 1,450 r.p.m., the vertical shaft at the forward end of the motor turns at 2,175 r.p.m., this being the speed of the two magnetos and the water pump. The lower vertical shaft bevel gear and the magneto driving gear are made integral with the vertical driving shaft, which is carried in plain bearings in an aluminium housing. This housing is clamped to the upper half of the crank case by means of three studs. The cam-shaft carries eighteen cams, these being the inlet and exhaust cams, and a set of half compression cams which are formed with the exhaust cams and are put into action when required by means of a lever at the forward end of the cam-shaft. The cam-shaft is hollow, and serves as a channel for the conveyance of lubricating oil to each of the camshaft bearings. At the forward end of this shaft there is also mounted an air pump for maintaining pressure on the fuel supply tank, and a bevel gear tachometer drive.

Lubrication of the engine is carried out by a full pressure system. The oil is pumped through a single manifold, with seven branches to the crankshaft main bearings, and then in turn through the hollow crankshaft to the connecting-rod big ends and thence through small tubes, already noted, to the small end bearings. The oil pump has four pistons and two double valves driven from a single eccentric shaft on which are mounted four eccentrics. The pump is continuously submerged in oil; in order to avoid great variations in pressure in the oil lines there is a piston operated pressure regulator, cut in between the pump and the oil lines. The two small pistons of the pump take fresh oil from a tank located in the fuselage of the machine; one of these delivers oil to the cam shaft, and one delivers to the crankshaft; this fresh oil mixes with the used oil, returns to the base, and back to the main large oil pump cylinders. By means of these small pump pistons a constant quantity of oil is kept in the motor, and the oil is continually being freshened by means of the new oil coming in. All the oil pipes are very securely fastened to the lower half of the crank case, and some cooling of the oil is effected by air passing through channels cast in the crank case on its way to the carburettor.

A light steel manifold serves to connect the exhaust ports of the cylinders to the main exhaust pipe, which is inclined about 25 degrees from vertical and is arranged to give on to the atmosphere just over the top of the upper wing of the aeroplane.

As regards carburation, an automatic air valve surrounds the throat of the carburettor, maintaining normal composition of mixture. A small jet is fitted for starting and running without load. The channels cast in the crank chamber, already alluded to in connection with oil-cooling, serve to warm the air before it reaches the carburettor, of which the body is water-jacketed.

Ignition of the engine is by means of two Bosch ZH6 magnetos, driven at a speed of 2,175 revolutions per minute when the engine is running at its normal speed of 1,450 revolutions. The maximum advance of spark is 12 mm., or 32 degrees before the top dead centre, and the firing order of the cylinders is 1,5,3,6,2,4.

The radiator fitted to this engine, together with the water-jackets, has a capacity of 25 litres of water, it is rectangular in shape, and is normally tilted at an angle of 30 degrees from vertical. Its weight is 26 kg., and it offers but slight head resistance in flight.

The radial type of engine, neglected altogether in Germany, was brought to a very high state of perfection at the end of the War period by British makers. Two makes, the Cosmos Engineering Company's 'Jupiter' and 'Lucifer,' and the A.B.C. 'Wasp II' and 'Dragon Fly 1A' require special mention for their light weight and reliability on trials.

The Cosmos 'Jupiter' was--for it is no longer being made--a 450 horse-power nine-cylinder radial engine, air-cooled, with the cylinders set in one single row; it was made both geared to reduce the propeller revolutions relatively to the crankshaft revolutions, and ungeared; the normal power of the geared type was 450 horse-power, and the total weight of the engine, including carburettors, magnetos, etc., was only 757 lbs.; the engine speed was 1,850 revolutions per minute, and the propeller revolutions were reduced by the gearing to 1,200. Fitted to a 'Bristol Badger' aeroplane, the total weight was 2,800 lbs., including pilot, passenger, two machine-guns, and full military load; at 7,000 feet the registered speed, with corrections for density, was 137 miles per hour; in climbing, the first 2,000 feet was accomplished in 1 minute 4 seconds; 4,000 feet was reached in 2 minutes 10 seconds; 6,000 feet was reached in 3 minutes 33 seconds, and 7,000 feet in 4 minutes 15 seconds. It was intended to modify the plane design and fit a new propeller, in order to attain even better results, but, if trials were made with these modifications, the results are not obtainable.

The Cosmos 'Lucifer' was a three-cylinder radial type engine of 100 horse-power, inverted Y design, made on the simplest possible principles with a view to quantity production and extreme reliability. The rated 100 horse-power was attained at 1,600 revolutions per minute, and the cylinder dimensions were 5.75 bore by 6.25 inches stroke. The cylinders were of aluminium and steel mixture, with aluminium heads; overhead valves, operated by push rods on the front side of the cylinders, were fitted, and a simple reducing gear ran them at half engine speed. The crank case was a circular aluminium casting, the engine being attached to the fuselage of the aeroplane by a circular flange situated at the back of the case; propeller shaft and crankshaft were integral. Dual ignition was provided, the generator and distributors being driven off the back end of the engine and the distributors being easily accessible. Lubrication was by means of two pumps, one scavenging and one suction, oil being fed under pressure from the crankshaft. A single carburettor fed all three cylinders, the branch pipe from the carburettor to the circular ring being provided with an exhaust heater. The total weight of the engine, 'all on,' was 280 lbs.

The A.B.C. 'Wasp II,' made by Walton Motors, Limited, is a seven-cylinder radial, air-cooled engine, the cylinders having a bore of 4.75 inches and stroke 6.25 inches. The normal brake horse-power at 1,650 revolutions is 160, and the maximum 200 at a speed of 1,850 revolutions per minute. Lubrication is by means of two rotary pumps, one feeding through the hollow crankshaft to the crank pin, giving centrifugal feed to big end and thence splash oiling, and one feeding to the nose of the engine, dropping on to the cams and forming a permanent sump for the gears on the bottom of the engine nose. Two carburettors are fitted, and two two-spark magnetos, running at one and three-quarters engine speed. The total weight of this engine is 350 lbs., or 1.75 lbs. per horse-power. Oil consumption at 1,850 revolutions is .03 pints per horse-power per hour, and petrol consumption is .56 pints per horsepower per hour. The engine thus shows as very economical in consumption, as well as very light in weight.

The A.B.C. 'Dragon Fly 1A' is a nine-cylinder radial engine having one overhead inlet and two overhead exhaust valves per cylinder. The cylinder dimensions are 5.5 inches bore by 6.5 inches stroke, and the normal rate of speed, 1,650 revolutions per minute, gives 340 horse-power. The oiling is by means of two pumps, the system being practically identical with that of the 'Wasp II.' Oil consumption is .021 pints per brake horse-power per hour, and petrol consumption .56 pints--the same as that of the 'Wasp II.' The weight of the complete engine, including propeller boss, is 600 lbs., or 1,765 lbs. per horse-power.

These A.B.C. radials have proved highly satisfactory on tests, and their extreme simplicity of design and reliability commend them as engineering products and at the same time demonstrate the value, for aero work, of the air-cooled radial design--when this latter is accompanied by sound workmanship. These and the Cosmos engines represent the minimum of weight per horse-power yet attained, together with a practicable degree of reliability, in radial and probably any aero engine design.

APPENDIX A
GENERAL MENSIER'S REPORT ON THE TRIALS OF CLEMENT ADER'S AVION.

Paris, October 21, 1897.

Report on the trials of M. Clement Ader's aviation apparatus.

M. Ader having notified the Minister of War by letter, July 21, 1897, that the Apparatus of Aviation which he had agreed to build under the conditions set forth in the convention of July 24th, 1894, was ready, and therefore requesting that trials be undertaken before a Committee appointed for this purpose as per the decision of August 4th, the Committee was appointed as follows:--

Division General Mensier, Chairman; Division General Delambre, Inspector General of the Permanent Works of Coast Defence, Member of the Technical Committee of the Engineering Corps; Colonel Laussedat, Director of the Conservatoire des Arts et Metiers; Sarrau, Member of the Institute, Professor of Mechanical Engineering at the Polytechnic School; Leaute, Member of the Institute, Professor of Mechanical Engineering at the Polytechnique School.

Colonel Laussedat gave notice at once that his health and work as Director of the Conservatoire des Arts et Metiers did not permit him to be a member of the Committee; the Minister therefore accepted his resignation on September 24th, and decided not to replace him.

Later on, however, on the request of the Chairman of the Committee, the Minister appointed a new member General Grillon, commanding the Engineer Corps of the Military Government of Paris.

To carry on the trials which were to take place at the camp of Satory, the Minister ordered the Governor of the Military Forces of Paris to requisition from the Engineer Corps, on the request of the Chairman of the Committee, the men necessary to prepare the grounds at Satory.

After an inspection made on the 16th an aerodrome was chosen. M. Ader's idea was to have it of circular shape with a width of 40 metres and an average diameter of 450 metres. The preliminary work, laying out the grounds, interior and exterior circumference, etc., was finished at the end of August; the work of smoothing off the grounds began September 1st with forty-five men and two rollers, and was finished on the day of the first tests, October 12th.

The first meeting of the Committee was held August 18th in M. Ader's workshop; the object being to demonstrate the machine to the Committee and give all the information possible on the tests that were to be held. After a careful examination and after having heard all the explanations by the inventor which were deemed useful and necessary, the Committee decided that the apparatus seemed to be built with a perfect understanding of the purpose to be fulfilled as far as one could judge from a study of the apparatus at rest; they therefore authorised M. Ader to take the machine apart and carry it to the camp at Satory so as to proceed with the trials.

By letter of August 19th the Chairman made report to the Minister of the findings of the Committee.

The work on the grounds having taken longer than was anticipated, the Chairman took advantage of this delay to call the Committee together for a second meeting, during which M. Ader was to run the two propulsive screws situated at the forward end of the apparatus.

The meeting was held October 2nd. It gave the Committee an opportunity to appreciate the motive power in all its details; firebox, boiler, engine, under perfect control, absolute condensation, automatic fuel and feed of the liquid to be vaporised, automatic lubrication and scavenging; everything, in a word, seemed well designed and executed.

The weights in comparison with the power of the engine realised a considerable advance over anything made to date, since the two engines weighed together realised 42 kg., the firebox and boiler 60 kg., the condenser 15 kg., or a total of 117 kg. for approximately 40 horse-power or a little less than 3 kg. per horse-power.

One of the members summed up the general opinion by saying: 'Whatever may be the result from an aviation point of view, a result which could not be foreseen for the moment, it was nevertheless proven that from a mechanical point of view M. Ader's apparatus was of the greatest interest and real ingeniosity. He expressed a hope that in any case the machine would not be lost to science.'

The second experiment in the workshop was made in the presence of the Chairman, the purpose being to demonstrate that the wings, having a spread of 17 metres, were sufficiently strong to support the weight of the apparatus. With this object in view, 14 sliding supports were placed under each one of these, representing imperfectly the manner in which the wings would support the machine in the air; by gradually raising the supports with the slides, the wheels on which the machine rested were lifted from the ground. It was evident at that time that the members composing the skeleton of the wings supported the apparatus, and it was quite evident that when the wings were supported by the air on every point of their surface, the stress would be better equalised than when resting on a few supports, and therefore the resistance to breakage would be considerably greater.

After this last test, the work on the ground being practically finished, the machine was transported to Satory, assembled and again made ready for trial.

At first M. Ader was to manoeuvre the machine on the ground at a moderate speed, then increase this until it was possible to judge whether there was a tendency for the machine to rise; and it was only after M. Ader had acquired sufficient practice that a meeting of the Committee was to be called to be present at the first part of the trials; namely, volutions of the apparatus on the ground.

The first test took place on Tuesday, October 12th, in the presence of the Chairman of the Committee. It had rained a good deal during the night and the clay track would have offered considerable resistance to the rolling of the machine; furthermore, a moderate wind was blowing from the south-west, too strong during the early part of the afternoon to allow of any trials.

Toward sunset, however, the wind having weakened, M. Ader decided to make his first trial; the machine was taken out of its hangar, the wings were mounted and steam raised. M. Ader in his seat had, on each side of him, one man to the right and one to the left, whose duty was to rectify the direction of the apparatus in the event that the action of the rear wheel as a rudder would not be sufficient to hold the machine in a straight course.

At 5.25 p.m. the machine was started, at first slowly and then at an increased speed; after 250 or 300 metres, the two men who were being dragged by the apparatus were exhausted and forced to fall flat on the ground in order to allow the wings to pass over them, and the trip around the track was completed, a total of 1,400 metres, without incident, at a fair speed, which could be estimated to be from 300 to 400 metres per minute. Notwithstanding M. Ader's inexperience, this being the first time that he had run his apparatus, he followed approximately the chalk line which marked the centre of the track and he stopped at the exact point from which he started.

The marks of the wheels on the ground, which was rather soft, did not show up very much, and it was clear that a part of the weight of the apparatus had been supported by the wings, though the speed was only about one-third of what the machine could do had M. Ader used all its motive power; he was running at a pressure of from 3 to 4 atmospheres, when he could have used 10 to 12.

This first trial, so fortunately accomplished, was of great importance; it was the first time that a comparatively heavy vehicle (nearly 400 kg., including the weight of the operator, fuel, and water) had been set in motion by a tractive apparatus, using the air solely as a propelling medium. The favourable report turned in by the Committee after the meeting of October 2nd was found justified by the results demonstrated on the grounds, and the first problem of aviation, namely, the creation of efficient motive power, could be considered as solved, since the propulsion of the apparatus in the air would be a great deal easier than the traction on the ground, provided that the second part of the problem, the sustaining of the machine in the air, would be realised.

The next day, Wednesday the 13th, no further trials were made on account of the rain and wind.

On Thursday the 14th the Chairman requested that General Grillon, who had just been appointed a member of the Committee, accompany him so as to have a second witness.

The weather was fine, but a fairly strong, gusty wind was blowing from the south. M. Ader explained to the two members of the Committee the danger of these gusts, since at two points of the circumference the wind would strike him sideways. The wind was blowing in the direction A B, the apparatus starting from C, and running in the direction shown by the arrow. The first dangerous spot would be at B. The apparatus had been kept in readiness in the event of the wind dying down. Toward sunset the wind seemed to die down, as it had done on the evening of the

12th. M. Ader hesitated, which, unfortunately, further events only justified, but decided to make a new trial.

At the start, which took place at 5.15 p.m., the apparatus, having the wind in the rear, seemed to run at a fairly regular speed; it was, nevertheless, easy to note from the marks of the wheels on the ground that the rear part of the apparatus had been lifted and that the rear wheel, being the rudder, had not been in constant contact with the ground. When the machine came to the neighbourhood of B, the two members of the Committee saw the machine swerve suddenly out of the track in a semicircle, lean over to the right and finally stop. They immediately proceeded to the point where the accident had taken place and endeavoured to find an explanation for the same. The Chairman finally decided as follows:

M. Ader was the victim of a gust of wind which he had feared as he explained before starting out; feeling himself thrown out of his course, he tried to use the rudder energetically, but at that time the rear wheel was not in contact with the ground, and therefore did not perform its function; the canvas rudder, which had as its purpose the manoeuvring of the machine in the air, did not have sufficient action on the ground. It would have been possible without any doubt to react by using the propellers at unequal speed, but M. Ader, being still inexperienced, had not thought of this. Furthermore, he was thrown out of his course so quickly that he decided, in order to avoid a more serious accident, to stop both engines. This sudden stop produced the half-circle already described and the fall of the machine on its side.

The damage to the machine was serious; consisting at first sight of the rupture of both propellers, the rear left wheel and the bending of the left wing tip. It will only be possible to determine after the machine is taken apart whether the engine, and more particularly the organs of transmission, have been put out of line.

Whatever the damage may be, though comparatively easy to repair, it will take a certain amount of time, and taking into consideration the time of year it is evident that the tests will have to be adjourned for the present.

As has been said in the above report, the tests, though prematurely interrupted, have shown results of great importance, and though the final results are hard to foresee, it would seem advisable to continue the trials. By waiting for the return of spring there will be plenty of time to finish the tests and it will not be necessary to rush matters, which was a partial cause of the accident. The Chairman of the Committee personally has but one hope, and that is that a decision be reached accordingly.

Division General,
    Chairman of the Committee,
        Mensier.
Boulogne-sur-Seine, October 21st, 1897.

Annex to the Report of October 21st.

General Grillon, who was present at the trials of the 14th, and who saw the report relative to what happened during that day, made the following observations in writing, which are reproduced herewith in quotation marks. The Chairman of the Committee does not agree with General Grillon and he answers these observations paragraph by paragraph.

1. 'If the rear wheel (there is only one of these) left but intermittent tracks on the ground, does that prove that the machine has a tendency to rise when running at a certain speed?'

Answer.--This does not prove anything in any way, and I was very careful not to mention this in my report, this point being exactly what was needed and that was not demonstrated during the two tests made on the grounds.

'Does not this unequal pressure of the two pair of wheels on the ground show that the centre of gravity of the apparatus is placed too far forward and that under the impulse of the propellers the machine has a tendency to tilt forward, due to the resistance of the air?'

Answer.--The tendency of the apparatus to rise from the rear when it was running with the wind seemed to be brought about by the effects of the wind on the huge wings, having a spread of 17 metres, and I believe that when the machine would have faced the wind the front wheels would have been lifted.

During the trials of October 12th, when a complete circuit of the track was accomplished without incidents, as I and Lieut. Binet witnessed, there was practically no wind. I was therefore unable to verify whether during this circuit the two front wheels or the rear wheel were in constant contact with the ground, because when the trial was over it was dark (it was 5.30) and the next day it was impossible to see anything because it had rained during the night and during Wednesday morning. But what would prove that the rear wheel was in contact with the ground at all times is the fact that M. Ader, though inexperienced, did not swerve from the circular track, which would prove that he steered pretty well with his rear wheel--this he could not have done if he had been in the air.

In the tests of the 12th, the speed was at least as great as on the 14th.

2. 'It would seem to me that if M. Ader thought that his rear wheels were off the ground he should have used his canvas rudder in order to regain his proper course; this was the best way of causing the machine to rotate, since it would have given an angular motion to the front axle.'

Answer.--I state in my report that the canvas rudder whose object was the manoeuvre of the apparatus in the air could have no effect on the apparatus on the ground, and to convince oneself of this point it is only necessary to consider the small surface of this canvas rudder compared with the mass to be handled on the ground, a weight of approximately 400 kg. According to my idea, and as I have stated in my report, M. Ader should have steered by increasing the speed on one of his propellers and slowing down the other. He admitted afterward that this remark was well founded, but that he did not have time to think of it owing to the suddenness of the accident.

3. 'When the apparatus fell on its side it was under the sole influence of the wind, since M. Ader had stopped the machine. Have we not a result here which will always be the same when the machine comes to the ground, since the engines will always have to be stopped or slowed down when coming to the ground? Here seems to be a bad defect of the apparatus under trial.'

Answer.--I believe that the apparatus fell on its side after coming to a stop, not on account of the wind, but because the semicircle described was on rough ground and one of the wheels had collapsed.

Mensier.
October 27th, 1897.

APPENDIX B

Specification and Claims of Wright Patent, No. 821393. Filed March 23rd, 1903. Issued May 22nd, 1906. Expires May 22nd, 1923.

To all whom it may concern.

Be it known that we, Orville Wright and Wilbur Wright, citizens of the United States, residing in the city of Dayton, county of Montgomery, and State of Ohio, have invented certain new and useful Improvements in Flying Machines, of which the following is a specification.

Our invention relates to that class of flying-machines in which the weight is sustained by the reactions resulting when one or more aeroplanes are moved through the air edgewise at a small angle of incidence, either by the application of mechanical power or by the utilisation of the force of gravity.

The objects of our invention are to provide means for maintaining or restoring the equilibrium or lateral balance of the apparatus, to provide means for guiding the machine both vertically and horizontally, and to provide a structure combining lightness, strength, convenience of construction and certain other advantages which will hereinafter appear.

To these ends our invention consists in certain novel features, which we will now proceed to describe and will then particularly point out in the claims. In the accompanying drawings, Figure I 1 is a perspective view of an apparatus embodying our invention in one form. Fig. 2 is a plan view of the same, partly in horizontal section and partly broken away. Fig. 3 is a side elevation, and Figs. 4 and 5 are detail views, of one form of flexible joint for connecting the upright standards with the aeroplanes.

In flying machines of the character to which this invention relates the apparatus is supported in the air by reason of the contact between the air and the under surface of one or more aeroplanes, the contact surface being presented at a small angle of incidence to the air. The relative movements of the air and aeroplane may be derived from the motion of the air in the form of wind blowing in the direction opposite to that in which the apparatus is travelling or by a combined downward and forward movement of the machine, as in starting from an elevated position or by combination of these two things, and in either case the operation is that of a soaring-machine, while power applied to the machine to propel it positively forward will cause the air to support the machine in a similar manner. In either case owing to the varying conditions to be met there are numerous disturbing forces which tend to shift the machine from the position which it should occupy to obtain the desired results. It is the chief object of our invention to provide means for remedying this difficulty, and we will now proceed to describe the construction by means of which these results are accomplished.

In the accompanying drawing we have shown an apparatus embodying our invention in one form. In this illustrative embodiment the machine is shown as comprising two parallel superposed aeroplanes, 1 and 2, may be embodied in a structure having a single aeroplane. Each aeroplane is of considerably greater width from side to side than from front to rear. The four corners of the upper aeroplane are indicated by the reference letters a, b, c, and d, while the corresponding corners of the lower aeroplane 2 are indicated by the reference letters e, f, g, and h. The marginal lines ab and ef indicate the front edges of the aeroplanes, the lateral margins of the upper aeroplane are indicated, respectively, by the lines ad and bc, the lateral margins of the lower aeroplane are indicated, respectively, by the lines eh and fg, while the rear margins of the upper and lower aeroplanes are indicated, respectively, by the lines cd and gh.

Before proceeding to a description of the fundamental theory of operation of the structure we will first describe the preferred mode of constructing the aeroplanes and those portions of the structure which serve to connect the two aeroplanes.

Each aeroplane is formed by stretching cloth or other suitable fabric over a frame composed of two parallel transverse spars 3, extending from side to side of the machine, their ends being connected by bows 4 extending from front to rear of the machine. The front and rear spars 3 of each aeroplane are connected by a series of parallel ribs 5, which preferably extend somewhat beyond the rear spar, as shown. These spars, bows, and ribs are preferably constructed of wood having the necessary strength, combined with lightness and flexibility. Upon this framework the cloth which forms the supporting surface of the aeroplane is secured, the frame being enclosed in the cloth. The cloth for each aeroplane previous to its attachment to its frame is cut on the bias and made up into a single piece approximately the size and shape of the aeroplane, having the threads of the fabric arranged diagonally to the transverse spars and longitudinal ribs, as indicated at 6 in Fig. 2. Thus the diagonal threads of the cloth form truss systems with the spars and ribs, the threads constituting the diagonal members. A hem is formed at the rear edge of the cloth to receive a wire 7, which is connected to the ends of the rear spar and supported by the rearwardly-extending ends of the longitudinal ribs 5, thus forming a rearwardly-extending flap or portion of the aeroplane. This construction of the aeroplane gives a surface which has very great strength to withstand lateral and longitudinal strains, at the same time being capable of being bent or twisted in the manner hereinafter described.

When two aeroplanes are employed, as in the construction illustrated, they are connected together by upright standards 8. These standards are substantially rigid, being preferably constructed of wood and of equal length, equally spaced along the front and rear edges of the aeroplane, to which they are connected at their top and bottom ends by hinged joints or universal joints of any suitable description. We have shown one form of connection which may be used for this purpose in Figs. 4 and 5 of the drawings. In this construction each end of the standard 8 has secured to it an eye 9 which engages with a hook 10, secured to a bracket plate 11, which latter plate is in turn fastened to the spar 3. Diagonal braces or stay-wires 12 extend from each end of each standard to the opposite ends of the adjacent standards, and as a convenient mode of attaching these parts I have shown a hook 13 made integral with the hook 10 to receive the end of one of the stay-wires, the other stay-wire being mounted on the hook 10. The hook 13 is shown as bent down to retain the stay-wire in connection to it, while the hook 10 is shown as provided with a pin 14 to hold the staywire 12 and eye 9 in position thereon. It will be seen that this construction forms a truss system which gives the whole machine great transverse rigidity and strength, while at the same time the jointed connections of the parts permit the aeroplanes to be bent or twisted in the manner which we will now proceed to describe.

15 indicates a rope or other flexible connection extending lengthwise of the

front of the machine above the lower aeroplane, passing under pulleys or other suitable guides 16 at the front corners e and f of the lower aeroplane, and extending thence upward and rearward to the upper rear corners c and d, of the upper aeroplane, where they are attached, as indicated at 17. To the central portion of the rope there is connected a laterally-movable cradle 18, which forms a means for moving the rope lengthwise in one direction or the other, the cradle being movable toward either side of the machine. We have devised this cradle as a convenient means for operating the rope 15, and the machine is intended to be generally used with the operator lying face downward on the lower aeroplane, with his head to the front, so that the operator's body rests on the cradle, and the cradle can be moved laterally by the movements of the operator's body. It will be understood, however, that the rope 15 may be manipulated in any suitable manner.

19 indicates a second rope extending transversely of the machine along the rear edge of the body portion of the lower aeroplane, passing under suitable pulleys or guides 20 at the rear corners g and h of the lower aeroplane and extending thence diagonally upward to the front corners a and b of the upper aeroplane, where its ends are secured in any suitable manner, as indicated at 21.

Considering the structure so far as we have now described it, and assuming that the cradle 18 be moved to the right in Figs. 1 and 2, as indicated by the arrows applied to the cradle in Fig. 1 and by the dotted lines in Fig. 2, it will be seen that that portion of the rope 15 passing under the guide pulley at the corner e and secured to the corner d will be under tension, while slack is paid out throughout the other side or half of the rope 15. The part of the rope 15 under tension exercises a downward pull upon the rear upper corner d of the structure and an upward pull upon the front lower corner e, as indicated by the arrows. This causes the corner d to move downward and the corner e to move upward. As the corner e moves upward it carries the corner a upward with it, since the intermediate standard 8 is substantially rigid and maintains an equal distance between the corners a and e at all times. Similarly, the standard 8, connecting the corners d and h, causes the corner h to move downward in unison with the corner d. Since the corner a thus moves upward and the corner h moves downward, that portion of the rope 19 connected to the corner a will be pulled upward through the pulley 20 at the corner h, and the pull thus exerted on the rope 19 will pull the corner b on the other wise of the machine downward and at the same time pull the corner g at said other side of the machine upward. This results in a downward movement of the corner b and an upward movement of the corner c. Thus it results from a lateral movement of the cradle 18 to the right in Fig. 1 that the lateral margins ad and eh at one side of the machine are moved from their normal positions in which they lie in the normal planes of their respective aeroplanes, into angular relations with said normal planes, each lateral margin on this side of the machine being raised above said normal plane at its forward end and depressed below said normal plane at its rear end, said lateral margins being thus inclined upward and forward. At the same time a reverse inclination is imparted to the lateral margins bc end fg at the other side of the machine, their inclination being downward and forward. These positions are indicated in dotted lines in Fig. 1 of the drawings. A movement of the cradle 18 in the opposite direction from its normal position will reverse the angular inclination of the lateral margins of the aeroplanes in an obvious manner. By reason of this construction it will be seen that with the particular mode of construction now under consideration it is possible to move the forward corner of the lateral edges of the aeroplane on one side of the machine either above or below the normal planes of the aeroplanes, a reverse movement of the forward corners of the lateral margins on the other side of the machine occurring simultaneously. During this operation each aeroplane is twisted or distorted around a line extending centrally across the same from the middle of one lateral margin to the middle of the other lateral margin, the twist due to the moving of the lateral margins to different angles extending across each aeroplane from side to side, so that each aeroplane surface is given a helicoidal warp or twist. We prefer this construction and mode of operation for the reason that it gives a gradually increasing angle to the body of each aeroplane from the centre longitudinal line thereof outward to the margin, thus giving a continuous surface on each side of the machine, which has a gradually increasing or decreasing angle of incidence from the centre of the machine to either side. We wish it to be understood, however, that our invention is not limited to this particular construction, since any construction whereby the angular relations of the lateral margins of the aeroplanes may be varied in opposite directions with respect to the normal planes of said aeroplanes comes within the scope of our invention. Furthermore, it should be understood that while the lateral margins of the aeroplanes move to different angular positions with respect to or above and below the normal planes of said aeroplanes, it does not necessarily follow that these movements bring the opposite lateral edges to different angles respectively above and below a horizontal plane since the normal planes of the bodies of the aeroplanes are inclined to the horizontal when the machine is in flight, said inclination being downward from front to rear, and while the forward corners on one side of the machine may be depressed below the normal planes of the bodies of the aeroplanes said depression is not necessarily sufficient to carry them below the horizontal planes passing through the rear corners on that side. Moreover, although we prefer to so construct the apparatus that the movements of the lateral margins on the opposite sides of the machine are equal in extent and opposite m direction, yet our invention is not limited to a construction producing this result, since it may be desirable under certain circumstances to move the lateral margins on one side of the machine just described without moving the lateral margins on the other side of the machine to an equal extent in the opposite direction. Turning now to the purpose of this provision for moving the lateral margins of the aeroplanes in the manner described, it should be premised that owing to various conditions of wind pressure and other causes the body of the machine is apt to become unbalanced laterally, one side tending to sink and the other side tending to rise, the machine turning around its central longitudinal axis. The provision which we have just described enables the operator to meet this difficulty and preserve the lateral balance of the machine. Assuming that for some cause that side of the machine which lies to the left of the observer in Figs. 1 and 2 has shown a tendency to drop downward, a movement of the cradle 18 to the right of said figures, as herein before assumed, will move the lateral margins of the aeroplanes in the manner already described, so that the margins ad and eh will be inclined downward and rearward, and the lateral margins bc and fg will be inclined upward and rearward with respect to the normal planes of the bodies of the aeroplanes. With the parts of the machine in this position it will be seen that the lateral margins ad and eh present a larger angle of incidence to the resisting air, while the lateral margins on the other side of the machine present a smaller angle of incidence. Owing to this fact, the side of the machine presenting the larger angle of incidence will tend to lift or move upward, and this upward movement will restore the lateral balance of the machine. When the other side of the machine tends to drop, a movement of the cradle 18 in the reverse direction will restore the machine to its normal lateral equilibrium. Of course, the same effect will be produced in the same way in the case of a machine employing only a single aeroplane.

In connection with the body of the machine as thus operated we employ a vertical rudder or tail 22, so supported as to turn around a vertical axis. This rudder is supported at the rear ends on supports or arms 23, pivoted at their forward ends to the rear margins of the upper and lower aeroplanes, respectively. These supports are preferably V-shaped, as shown, so that their forward ends are comparatively widely separated, their pivots being indicated at 24. Said supports are free to swing upward at their free rear ends, as indicated in dotted lines in Fig. 3, their downward movement being limited in any suitable manner. The vertical pivots of the rudder 22 are indicated at 25, and one of these pivots has mounted thereon a sheave or pulley 26, around which passes a tiller-rope 27, the ends of which are extended out laterally and secured to the rope 19 on opposite sides of the central point of said rope. By reason of this construction the lateral shifting of the cradle 18 serves to turn the rudder to one side or the other of the line of flight. It will be observed in this connection that the construction is such that the rudder will always be so turned as to present its resisting surface on that side of the machine on which the lateral margins of the aeroplanes present the least angle of resistance. The reason of this construction is that when the lateral margins of the aeroplanes are so turned in the manner hereinbefore described as to present different angles of incidence to the atmosphere, that side presenting the largest angle of incidence, although being lifted or moved upward in the manner already described, at the same time meets with an increased resistance to its forward motion, while at the same time the other side of the machine, presenting a smaller angle of incidence, meets with less resistance to its forward motion and tends to move forward more rapidly than the retarded side. This gives the machine a tendency to turn around its vertical axis, and this tendency if not properly met will not only change the direction of the front of the machine, but will ultimately permit one side thereof to drop into a position vertically below the other side with the aero planes in vertical position, thus causing the machine to fall. The movement of the rudder, hereinbefore described, prevents this action, since it exerts a retarding influence on that side of the machine which tends to move forward too rapidly and keeps the machine with its front properly presented to the direction of flight and with its body properly balanced around its central longitudinal axis. The pivoting of the supports 23 so as to permit them to swing upward prevents injury to the rudder and its supports in case the machine alights at such an angle as to cause the rudder to strike the ground first, the parts yielding upward, as indicated in dotted lines in Fig. 3, and thus preventing injury or breakage. We wish it to be understood, however, that we do not limit ourselves to the particular description of rudder set forth, the essential being that the rudder shall be vertical and shall be so moved as to present its resisting surface on that side of the machine which offers the least resistance to the atmosphere, so as to counteract the tendency of the machine to turn around a vertical axis when the two sides thereof offer different resistances to the air.

From the central portion of the front of the machine struts 28 extend horizontally forward from the lower aeroplane, and struts 29 extend downward and forward from the central portion of the upper aeroplane, their front ends being united to the struts 28, the forward extremities of which are turned up, as indicated at 30. These struts 28 and 29 form truss-skids projecting in front of the whole frame of the machine and serving to prevent the machine from rolling over forward when it alights. The struts 29 serve to brace the upper portion of the main frame and resist its tendency to move forward after the lower aeroplane has been stopped by its contact with the earth, thereby relieving the rope 19 from undue strain, for it will be understood that when the machine comes into contact with the earth, further forward movement of the lower portion thereof being suddenly arrested, the inertia of the upper portion would tend to cause it to continue to move forward if not prevented by the struts 29, and this forward movement of the upper portion would bring a very violent strain upon the rope 19, since it is fastened to the upper portion at both of its ends, while its lower portion is connected by the guides 20 to the lower portion. The struts 28 and 29 also serve to support the front or horizontal rudder, the construction of which we will now proceed to describe.

The front rudder 31 is a horizontal rudder having a flexible body, the same consisting of three stiff crosspieces or sticks 32, 33, and 34, and the flexible ribs 35, connecting said cross-pieces and extending from front to rear. The frame thus provided is covered by a suitable fabric stretched over the same to form the body of the rudder. The rudder is supported from the struts 29 by means of the intermediate cross-piece 32, which is located near the centre of pressure slightly in front of a line equidistant between the front and rear edges of the rudder, the cross-piece 32 forming the pivotal axis of the rudder, so as to constitute a balanced rudder. To the front edge of the rudder there are connected springs 36 which springs are connected to the upturned ends 30 of the struts 28, the construction being such that said springs tend to resist any movement either upward or downward of the front edge of the horizontal rudder. The rear edge of the rudder lies immediately in front of the operator and may be operated by him in any suitable manner. We have shown a mechanism for this purpose comprising a roller or shaft 37, which may be grasped by the operator so as to turn the same in either direction. Bands 38 extend from the roller 37 forward to and around a similar roller or shaft 39, both rollers or shafts being supported in suitable bearings on the struts 28. The forward roller or shaft has rearwardly-extending arms 40, which are connected by links 41 with the rear edge of the rudder 31. The normal position of the rudder 31 is neutral or substantially parallel with the aeroplanes 1 and 2; but its rear edge may be moved upward or downward, so as to be above or below the normal plane of said rudder through the mechanism provided for that purpose. It will be seen that the springs 36 will resist any tendency of the forward edge of the rudder to move in either direction, so that when force is applied to the rear edge of said rudder the longitudinal ribs 35 bend, and the rudder thus presents a concave surface to the action of the wind either above or below its normal plane, said surface presenting a small angle of incidence at its forward portion and said angle of incidence rapidly increasing toward the rear. This greatly increases the efficiency of the rudder as compared with a plane surface of equal area. By regulating the pressure on the upper and lower sides of the rudder through changes of angle and curvature in the manner described a turning movement of the main structure around its transverse axis may be effected, and the course of the machine may thus be directed upward or downward at the will of the operator and the longitudinal balance thereof maintained.

Contrary to the usual custom, we place the horizontal rudder in front of the aeroplanes at a negative angle and employ no horizontal tail at all. By this arrangement we obtain a forward surface which is almost entirely free from pressure under ordinary conditions of flight, but which even if not moved at all from its original position becomes an efficient lifting-surface whenever the speed of the machine is accidentally reduced very much below the normal, and thus largely counteracts that backward travel of the centre of pressure on the aeroplanes which has frequently been productive of serious injuries by causing the machine to turn downward and forward and strike the ground head-on. We are aware that a forward horizontal rudder of different construction has been used in combination with a supporting surface and a rear horizontal-rudder; but this combination was not intended to effect and does not effect the object which we obtain by the arrangement hereinbefore described.

We have used the term 'aeroplane' in this specification and the appended claims to indicate the supporting surface or supporting surfaces by means of which the machine is sustained in the air, and by this term we wish to be understood as including any suitable supporting surface which normally is substantially flat, although. Of course, when constructed of cloth or other flexible fabric, as we prefer to construct them, these surfaces may receive more or less curvature from the resistance of the air, as indicated in Fig. 3.

We do not wish to be understood as limiting ourselves strictly to the precise details of construction hereinbefore described and shown in the accompanying drawings, as it is obvious that these details may be modified without departing from the principles of our invention. For instance, while we prefer the construction illustrated in which each aeroplane is given a twist along its entire length in order to set its opposite lateral margins at different angles, we have already pointed out that our invention is not limited to this form of construction, since it is only necessary to move the lateral marginal portions, and where these portions alone are moved only those upright standards which support the movable portion require flexible connections at their ends.

Having thus fully described our invention, what we claim as new, and desire to secure by Letters Patent, is:--

1. In a flying machine, a normally flat aeroplane having lateral marginal portions capable of movement to different positions above or below the normal plane of the body of the aeroplane, such movement being about an axis transverse to the line of flight, whereby said lateral marginal portions may be moved to different angles relatively to the normal plane of the body of the aeroplane, so as to present to the atmosphere different angles of incidence, and means for so moving said lateral

marginal portions, substantially as described.

2. In a flying machine, the combination, with two normally parallel aeroplanes, superposed the one above the other, of upright standards connecting said planes at their margins, the connections between the standards and aeroplanes at the lateral portions of the aeroplanes being by means of flexible joints, each of said aeroplanes having lateral marginal portions capable of movement to different positions above or below the normal plane of the body of the aeroplane, such movement being about an axis transverse to the line of flight, whereby said lateral marginal portions may be moved to different angles relatively to the normal plane of the body of the aeroplane, so as to present to the atmosphere different angles of incidence, the standards maintaining a fixed distance between the portions of the aeroplanes which they connect, and means for imparting such movement to the lateral marginal portions of the aeroplanes, substantially as described.

3. In a flying machine, a normally flat aeroplane having lateral marginal portions capable of movement to different positions above or below the normal plane of the body of the aeroplane, such movement being about an axis transverse to the line of flight, whereby said lateral marginal portions may be moved to different angles relatively to the normal plane of the body of the aeroplane, and also to different angles relatively to each other, so as to present to the atmosphere different angles of incidence, and means for simultaneously imparting such movement to said lateral marginal portions, substantially as described.

4. In a flying machine, the combination, with parallel superposed aeroplanes, each having lateral marginal portions capable of movement to different positions above or below the normal plane of the body of the aeroplane, such movement being about an axis transverse to the line of flight, whereby said lateral marginal portions may be moved to different angles relatively to the normal plane of the body of the aeroplane, and to different angles relatively to each other, so as to present to the atmosphere different angles of incidence, of uprights connecting said aeroplanes at their edges, the uprights connecting the lateral portions of the aeroplanes being connected with said aeroplanes by flexible joints, and means for simultaneously imparting such movement to said lateral marginal portions, the standards maintaining a fixed distance between the parts which they connect, whereby the lateral portions on the same side of the machine are moved to the same angle, substantially

as described.

5. In a flying machine, an aeroplane having substantially the form of a normally flat rectangle elongated transversely to the line of flight, in combination which means for imparting to the lateral margins of said aeroplane a movement about an axis lying in the body of the aeroplane perpendicular to said lateral margins, and thereby moving said lateral margins into different angular relations to the normal plane of the body of the aeroplane, substantially as described.

6. In a flying machine, the combination, with two superposed and normally parallel aeroplanes, each having substantially the form of a normally flat rectangle elongated transversely to the line of flight, of upright standards connecting the edges of said aeroplanes to maintain their equidistance, those standards at the lateral portions of said aeroplanes being connected therewith by flexible joints, and means for simultaneously imparting to both lateral margins of both aeroplanes a movement about axes which are perpendicular to said margins and in the planes of the bodies of the respective aeroplanes, and thereby moving the lateral margins on the opposite sides of the machine into different angular relations to the normal planes of the respective aeroplanes, the margins on the same side of the machine moving to the same angle, and the margins on one side of the machine moving to an angle different from the angle to which the margins on the other side of the machine move, substantially as described.

7. In a flying machine, the combination, with an aeroplane, and means for simultaneously moving the lateral portions thereof into different angular relations to the normal plane of the body of the aeroplane and to each other, so as to present to the atmosphere different angles of incidence, of a vertical rudder, and means whereby said rudder is caused to present to the wind that side thereof nearest the side of the aeroplane having the smaller angle of incidence and offering the least resistance to the atmosphere, substantially as described.

8. In a flying machine, the combination, with two superposed and normally parallel aeroplanes, upright standards connecting the edges of said aeroplanes to maintain their equidistance, those standards at the lateral portions of said aeroplanes being connected therewith by flexible joints, and means for simultaneously moving both lateral portions of both aeroplanes into different angular relations to the normal planes of the bodies of the respective aeroplanes, the lateral portions on

one side of the machine being moved to an angle different from that to which the lateral portions on the other side of the machine are moved, so as to present different angles of incidence at the two sides of the machine, of a vertical rudder, and means whereby said rudder is caused to present to the wind that side thereof nearest the side of the aeroplanes having the smaller angle of incidence and offering the least resistance to the atmosphere, substantially as described.

9. In a flying machine, an aeroplane normally flat and elongated transversely to the line of flight, in combination with means for imparting to said aeroplane a helicoidal warp around an axis transverse to the line of flight and extending centrally along the body aeroplane in the direction of the elongation aeroplane, substantially as described.

10. In a flying machine, two aeroplanes, each normally flat and elongated transversely to the line of flight, and upright standards connecting the edges of said aeroplanes to maintain their equidistance, the connections between said standards and aeroplanes being by means of flexible joints, in combination with means for simultaneously imparting to each of said aeroplanes a helicoidal warp around an axis transverse to the line of flight and extending centrally along the body of the aeroplane in the direction of the aeroplane, substantially as described.

11. In a flying machine, two aeroplanes, each normally flat and elongated transversely to the line of flight, and upright standards connecting the edges of said aeroplanes to maintain their equidistance, the connections between such standards and aeroplanes being by means of flexible joints, in combination with means for simultaneously imparting to each of said aeroplanes a helicoidal warp around an axis transverse to the line of flight and extending centrally along the body of the aeroplane in the direction of the elongation of the aeroplane, a vertical rudder, and means whereby said rudder is caused to present to the wind that side thereof nearest the side of the aeroplanes having the smaller angle of incidence and offering the least resistance to the atmosphere, substantially as described.

12. In a flying machine, the combination, with an aeroplane, of a normally flat and substantially horizontal flexible rudder, and means for curving said rudder rearwardly and upwardly or rearwardly and downwardly with respect to its normal plane, substantially as described.

13. In a flying machine, the combination, with an aeroplane, of a normally flat

and substantially horizontal flexible rudder pivotally mounted on an axis transverse to the line of flight near its centre, springs resisting vertical movement of the front edge of said rudder, and means for moving the rear edge of said rudder, above or below the normal plane thereof, substantially as described.

14. A flying machine comprising superposed connected aeroplanes means for moving the opposite lateral portions of said aeroplanes to different angles to the normal planes thereof, a vertical rudder, means for moving said vertical rudder toward that side of the machine presenting the smaller angle of incidence and the least resistance to the atmosphere, and a horizontal rudder provided with means for presenting its upper or under surface to the resistance of the atmosphere, substantially as described.

15. A flying machine comprising superposed connected aeroplanes, means for moving the opposite lateral portions of said aeroplanes to different angles to the normal planes thereof, a vertical rudder, means for moving said vertical rudder toward that side of the machine presenting the smaller angle of incidence and the least resistance to the atmosphere, and a horizontal rudder provided with means for presenting its upper or under surface to the resistance of the atmosphere, said vertical rudder being located at the rear of the machine and said horizontal rudder at the front of the machine, substantially as described.

16. In a flying machine, the combination, with two superposed and connected aeroplanes, of an arm extending rearward from each aeroplane, said arms being parallel and free to swing upward at their rear ends, and a vertical rudder pivotally mounted in the rear ends of said arms, substantially as described.

17. A flying machine comprising two superposed aeroplanes, normally flat but flexible, upright standards connecting the margins of said aeroplanes, said standards being connected to said aeroplanes by universal joints, diagonal stay-wires connecting the opposite ends of the adjacent standards, a rope extending along the front edge of the lower aeroplane, passing through guides at the front corners thereof, and having its ends secured to the rear corners of the upper aeroplane, and a rope extending along the rear edge of the lower aeroplane, passing through guides at the rear corners thereof, and having its ends secured to the front corners of the upper aeroplane, substantially as described.

18. A flying machine comprising two superposed aeroplanes, normally flat but

flexible, upright standards connecting the margins of said aeroplanes, said standards being connected to said aeroplanes by universal joints, diagonal stay-wires connecting the opposite ends of the adjacent standards, a rope extending along the front edge of the lower aeroplane, passing through guides at the front corners thereof, and having its ends secured to the rear corners of the upper aeroplane, and a rope extending along the rear edge of the lower aeroplane, passing through guides at the rear corners thereof, and having its ends secured to the front corners of the upper aeroplane, in combination with a vertical rudder, and a tiller-rope connecting said rudder with the rope extending along the rear edge of the lower aeroplane, substantially as described.

ORVILLE WRIGHT.
WILBUR WRIGHT.

Witnesses:
Chas. E. Taylor.
E. Earle Forrer.

APPENDIX C

Proclamation published by the French Government on balloon ascents, 1783.
NOTICE TO THE PUBLIC! PARIS, 27TH AUGUST, 1783.

On the Ascent of balloons or globes in the air. The one in question has been raised in Paris this day, 27th August, 1783, at 5 p.m., in the Champ de Mars.

A Discovery has been made, which the Government deems it right to make known, so that alarm be not occasioned to the people.

On calculating the different weights of hot air, hydrogen gas, and common air, it has been found that a balloon filled with either of the two former will rise toward heaven till it is in equilibrium with the surrounding air, which may not happen until it has attained a great height.

The first experiment was made at Annonay, in Vivarais, MM. Montgolfier, the inventors; a globe formed of canvas and paper, 105 feet in circumference, filled with heated air, reached an uncalculated height. The same experiment has just been renewed in Paris before a great crowd. A globe of taffetas or light canvas covered by elastic gum and filled with inflammable air, has risen from the Champ de Mars, and been lost to view in the clouds, being borne in a north-westerly direction. One cannot foresee where it will descend.

It is proposed to repeat these experiments on a larger scale. Any one who shall see in the sky such a globe, which resembles 'la lune obscurcie,' should be aware that, far from being an alarming phenomenon, it is only a machine that cannot possibly cause any harm, and which will some day prove serviceable to the wants of society.

DE SAUVIGNY.
LENOIR.

www.bookjungle.com email: sales@bookjungle.com fax: 630-214-0564 mail: Book Jungle PO Box 2226 Champaign, IL 61825

The Codes Of Hammurabi And Moses
W. W. Davies

QTY

The discovery of the Hammurabi Code is one of the greatest achievements of archaeology, and is of paramount interest, not only to the student of the Bible, but also to all those interested in ancient history...

Religion   **ISBN:** *1-59462-338-4*   Pages:132
MSRP $12.95

The Theory of Moral Sentiments
Adam Smith

QTY

This work from 1749. contains original theories of conscience amd moral judgment and it is the foundation for systemof morals.

Philosophy   **ISBN:** *1-59462-777-0*   Pages:536
MSRP $19.95

Jessica's First Prayer
Hesba Stretton

QTY

In a screened and secluded corner of one of the many railway-bridges which span the streets of London there could be seen a few years ago, from five o'clock every morning until half past eight, a tidily set-out coffee-stall, consisting of a trestle and board, upon which stood two large tin cans, with a small fire of charcoal burning under each so as to keep the coffee boiling during the early hours of the morning when the work-people were thronging into the city on their way to their daily toil...

Childrens   **ISBN:** *1-59462-373-2*   Pages:84
MSRP $9.95

My Life and Work
Henry Ford

QTY

Henry Ford revolutionized the world with his implementation of mass production for the Model T automobile. Gain valuable business insight into his life and work with his own auto-biography... "We have only started on our development of our country we have not as yet, with all our talk of wonderful progress, done more than scratch the surface. The progress has been wonderful enough but..."

Biographies/   **ISBN:** *1-59462-198-5*   Pages:300
MSRP $21.95

www.bookjungle.com *email: sales@bookjungle.com fax: 630-214-0564 mail: Book Jungle PO Box 2226 Champaign, IL 61825*

The Art of Cross-Examination
Francis Wellman

QTY

I presume it is the experience of every author, after his first book is published upon an important subject, to be almost overwhelmed with a wealth of ideas and illustrations which could readily have been included in his book, and which to his own mind, at least, seem to make a second edition inevitable. Such certainly was the case with me; and when the first edition had reached its sixth impression in five months, I rejoiced to learn that it seemed to my publishers that the book had met with a sufficiently favorable reception to justify a second and considerably enlarged edition. ..

Pages:412

Reference   ISBN: *1-59462-647-2*   *MSRP $19.95*

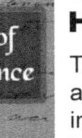

On the Duty of Civil Disobedience
Henry David Thoreau

QTY

Thoreau wrote his famous essay, On the Duty of Civil Disobedience, as a protest against an unjust but popular war and the immoral but popular institution of slave-owning. He did more than write—he declined to pay his taxes, and was hauled off to gaol in consequence. Who can say how much this refusal of his hastened the end of the war and of slavery ?

Law   ISBN: *1-59462-747-9*   **Pages:48**

MSRP $7.45

Dream Psychology Psychoanalysis for Beginners
Sigmund Freud

QTY

Sigmund Freud, born Sigismund Schlomo Freud (May 6, 1856 - September 23, 1939), was a Jewish-Austrian neurologist and psychiatrist who co-founded the psychoanalytic school of psychology. Freud is best known for his theories of the unconscious mind, especially involving the mechanism of repression; his redefinition of sexual desire as mobile and directed towards a wide variety of objects; and his therapeutic techniques, especially his understanding of transference in the therapeutic relationship and the presumed value of dreams as sources of insight into unconscious desires.

Pages:196

Psychology   ISBN: *1-59462-905-6*   *MSRP $15.45*

The Miracle of Right Thought
Orison Swett Marden

QTY

Believe with all of your heart that you will do what you were made to do. When the mind has once formed the habit of holding cheerful, happy, prosperous pictures, it will not be easy to form the opposite habit. It does not matter how improbable or how far away this realization may see, or how dark the prospects may be, if we visualize them as best we can, as vividly as possible, hold tenaciously to them and vigorously struggle to attain them, they will gradually become actualized, realized in the life. But a desire, a longing without endeavor, a yearning abandoned or held indifferently will vanish without realization.

Pages:360

Self Help   ISBN: *1-59462-644-8*   *MSRP $25.45*

www.bookjungle.com email: sales@bookjungle.com fax: 630-214-0564 mail: Book Jungle PO Box 2226 Champaign, IL 61825

QTY

- [ ] **The Rosicrucian Cosmo-Conception Mystic Christianity** by *Max Heindel* — ISBN: 1-59462-188-8 **$38.95**
The Rosicrucian Cosmo-conception is not dogmatic, neither does it appeal to any other authority than the reason of the student. It is: not controversial, but is: sent forth in the hope that it may help to clear...
New Age/Religion Pages 646

- [ ] **Abandonment To Divine Providence** by *Jean-Pierre de Caussade* — ISBN: 1-59462-228-0 **$25.95**
"The Rev. Jean Pierre de Caussade was one of the most remarkable spiritual writers of the Society of Jesus in France in the 18th Century. His death took place at Toulouse in 1751. His works have gone through many editions and have been republished...
Inspirational/Religion Pages 400

- [ ] **Mental Chemistry** by *Charles Haanel* — ISBN: 1-59462-192-6 **$23.95**
Mental Chemistry allows the change of material conditions by combining and appropriately utilizing the power of the mind. Much like applied chemistry creates something new and unique out of careful combinations of chemicals the mastery of mental chemistry...
New Age Pages 354

- [ ] **The Letters of Robert Browning and Elizabeth Barret Barrett 1845-1846 vol II** — ISBN: 1-59462-193-4 **$35.95**
by *Robert Browning* and *Elizabeth Barrett*
Biographies Pages 596

- [ ] **Gleanings In Genesis (volume I)** by *Arthur W. Pink* — ISBN: 1-59462-130-6 **$27.45**
Appropriately has Genesis been termed "the seed plot of the Bible" for in it we have, in germ form, almost all of the great doctrines which are afterwards fully developed in the books of Scripture which follow...
Religion/Inspirational Pages 420

- [ ] **The Master Key** by *L. W. de Laurence* — ISBN: 1-59462-001-6 **$30.95**
In no branch of human knowledge has there been a more lively increase of the spirit of research during the past few years than in the study of Psychology, Concentration and Mental Discipline. The requests for authentic lessons in Thought Control, Mental Discipline and...
New Age Pages 422

- [ ] **The Lesser Key Of Solomon Goetia** by *L. W. de Laurence* — ISBN: 1-59462-092-X **$9.95**
This translation of the first book of the "Lemegton" which is now for the first time made accessible to students of Talismanic Magic was done, after careful collation and edition, from numerous Ancient Manuscripts in Hebrew, Latin, and French...
New Age/Occult Pages 92

- [ ] **Rubaiyat Of Omar Khayyam** by *Edward Fitzgerald* — ISBN:1-59462-332-5 **$13.95**
Edward Fitzgerald, whom the world has already learned, in spite of his own efforts to remain within the shadow of anonymity, to look upon as one of the rarest poets of the century, was born at Bredfield, in Suffolk, on the 31st of March, 1809. He was the third son of John Purcell...
Music Pages 172

- [ ] **Ancient Law** by *Henry Maine* — ISBN: 1-59462-128-4 **$29.95**
The chief object of the following pages is to indicate some of the earliest ideas of mankind, as they are reflected in Ancient Law, and to point out the relation of those ideas to modern thought.
Religion/History Pages 452

- [ ] **Far-Away Stories** by *William J. Locke* — ISBN: 1-59462-129-2 **$19.45**
"Good wine needs no bush, but a collection of mixed vintages does. And this book is just such a collection. Some of the stories I do not want to remain buried for ever in the museum files of dead magazine-numbers an author's not unpardonable vanity..."
Fiction Pages 272

- [ ] **Life of David Crockett** by *David Crockett* — ISBN: 1-59462-250-7 **$27.45**
"Colonel David Crockett was one of the most remarkable men of the times in which he lived. Born in humble life, but gifted with a strong will, an indomitable courage, and unremitting perseverance...
Biographies/New Age Pages 424

- [ ] **Lip-Reading** by *Edward Nitchie* — ISBN: 1-59462-206-X **$25.95**
Edward B. Nitchie, founder of the New York School for the Hard of Hearing, now the Nitchie School of Lip-Reading, Inc, wrote "LIP-READING Principles and Practice". The development and perfecting of this meritorious work on lip-reading was an undertaking...
How-to Pages 400

- [ ] **A Handbook of Suggestive Therapeutics, Applied Hypnotism, Psychic Science** — ISBN: 1-59462-214-0 **$24.95**
by *Henry Munro*
Health/New Age/Health/Self-help Pages 376

- [ ] **A Doll's House: and Two Other Plays** by *Henrik Ibsen* — ISBN: 1-59462-112-8 **$19.95**
Henrik Ibsen created this classic when in revolutionary 1848 Rome. Introducing some striking concepts in playwriting for the realist genre, this play has been studied the world over.
Fiction/Classics/Plays 308

- [ ] **The Light of Asia** by *sir Edwin Arnold* — ISBN: 1-59462-204-3 **$13.95**
In this poetic masterpiece, Edwin Arnold describes the life and teachings of Buddha. The man who was to become known as Buddha to the world was born as Prince Gautama of India but he rejected the worldly riches and abandoned the reigns of power when...
Religion/History/Biographies Pages 170

- [ ] **The Complete Works of Guy de Maupassant** by *Guy de Maupassant* — ISBN: 1-59462-157-8 **$16.95**
"For days and days, nights and nights, I had dreamed of that first kiss which was to consecrate our engagement, and I knew not on what spot I should put my lips..."
Fiction/Classics Pages 240

- [ ] **The Art of Cross-Examination** by *Francis L. Wellman* — ISBN: 1-59462-309-0 **$26.95**
Written by a renowned trial lawyer, Wellman imparts his experience and uses case studies to explain how to use psychology to extract desired information through questioning.
How-to/Science/Reference Pages 408

- [ ] **Answered or Unanswered?** by *Louisa Vaughan* — ISBN: 1-59462-248-5 **$10.95**
Miracles of Faith in China
Religion Pages 112

- [ ] **The Edinburgh Lectures on Mental Science (1909)** by *Thomas* — ISBN: 1-59462-008-3 **$11.95**
This book contains the substance of a course of lectures recently given by the writer in the Queen Street Hall, Edinburgh. Its purpose is to indicate the Natural Principles governing the relation between Mental Action and Material Conditions...
New Age/Psychology Pages 148

- [ ] **Ayesha** by *H. Rider Haggard* — ISBN: 1-59462-301-5 **$24.95**
Verily and indeed it is the unexpected that happens! Probably if there was one person upon the earth from whom the Editor of this, and of a certain previous history, did not expect to hear again...
Classics Pages 380

- [ ] **Ayala's Angel** by *Anthony Trollope* — ISBN: 1-59462-352-X **$29.95**
The two girls were both pretty, but Lucy who was twenty-one who supposed to be simple and comparatively unattractive, whereas Ayala was credited, as her Bombwhat romantic name might show, with poetic charm and a taste for romance. Ayala when her father died was nineteen...
Fiction Pages 484

- [ ] **The American Commonwealth** by *James Bryce* — ISBN: 1-59462-286-8 **$34.45**
An interpretation of American democratic political theory. It examines political mechanics and society from the perspective of Scotsman James Bryce
Politics Pages 572

- [ ] **Stories of the Pilgrims** by *Margaret P. Pumphrey* — ISBN: 1-59462-116-0 **$17.95**
This book explores pilgrims religious oppression in England as well as their escape to Holland and eventual crossing to America on the Mayflower, and their early days in New England...
History Pages 268

www.bookjungle.com email: sales@bookjungle.com fax: 630-214-0564 mail: Book Jungle PO Box 2226 Champaign, IL 61825

			QTY
The Fasting Cure by *Sinclair Upton*	ISBN: *1-59462-222-1*	**$13.95**	☐
In the Cosmopolitan Magazine for May, 1910, and in the Contemporary Review (London) for April, 1910, I published an article dealing with my experiences in fasting. I have written a great many magazine articles, but never one which attracted so much attention... *New Age/Self Help/Health Pages 164*			
Hebrew Astrology by *Sepharial*	ISBN: *1-59462-308-2*	**$13.45**	☐
In these days of advanced thinking it is a matter of common observation that we have left many of the old landmarks behind and that we are now pressing forward to greater heights and to a wider horizon than that which represented the mind-content of our progenitors... *Astrology Pages 144*			
Thought Vibration or The Law of Attraction in the Thought World	ISBN: *1-59462-127-6*	**$12.95**	☐
by *William Walker Atkinson*	*Psychology/Religion Pages 144*		
Optimism by *Helen Keller*	ISBN: *1-59462-108-X*	**$15.95**	☐
Helen Keller was blind, deaf, and mute since 19 months old, yet famously learned how to overcome these handicaps, communicate with the world, and spread her lectures promoting optimism. An inspiring read for everyone... *Biographies/Inspirational Pages 84*			
Sara Crewe by *Frances Burnett*	ISBN: *1-59462-360-0*	**$9.45**	☐
In the first place, Miss Minchin lived in London. Her home was a large, dull, tall one, in a large, dull square, where all the houses were alike, and all the sparrows were alike, and where all the door-knockers made the same heavy sound... *Childrens/Classic Pages 88*			
The Autobiography of Benjamin Franklin by *Benjamin Franklin*	ISBN: *1-59462-135-7*	**$24.95**	☐
The Autobiography of Benjamin Franklin has probably been more extensively read than any other American historical work, and no other book of its kind has had such ups and downs of fortune. Franklin lived for many years in England, where he was agent... *Biographies/History Pages 332*			

Name	
Email	
Telephone	
Address	
City, State ZIP	

☐ Credit Card   ☐ Check / Money Order

Credit Card Number	
Expiration Date	
Signature	

Please Mail to: Book Jungle
                  PO Box 2226
                  Champaign, IL 61825
or Fax to: 630-214-0564

ORDERING INFORMATION

web: *www.bookjungle.com*
email: *sales@bookjungle.com*
fax: *630-214-0564*
mail: *Book Jungle PO Box 2226 Champaign, IL 61825*
or PayPal *to sales@bookjungle.com*

Please contact us for bulk discounts

DIRECT-ORDER TERMS

20% Discount if You Order Two or More Books
Free Domestic Shipping!
Accepted: Master Card, Visa, Discover, American Express

www.ingramcontent.com/pod-product-compliance
Lightning Source LLC
Chambersburg PA
CBHW082108230426
43671CB00015B/2637